BRUCELLOSIS IN THE GREATER YELLOWSTONE AREA

Norman F. Cheville, Principal Investigator
Dale R. McCullough, Principal Investigator
Lee R. Paulson, Project Director

Board on Agriculture
Board on Environmental Studies and Toxicology

Board on Agriculture
Commission on Life Sciences

National Research Council
Washington, D.C. 1998

NATIONAL ACADEMY PRESS   2101 Constitution Ave., N.W.   Washington, D.C. 20418

NOTICE: The project that is the subject of this report was approved by the Governing Board of the National Research Council, whose members are drawn from the councils of the National Academy of Sciences, the National Academy of Engineering, and the Institute of Medicine. The authors responsible for the report were chosen for their special competence and with regard for appropriate balance.

This report has been reviewed by a group other than the authors according to procedures approved by a Report Review Committee consisting of members of the National Academy of Sciences, the National Academy of Engineering, and the Institute of Medicine.

The National Academy of Sciences is a private, nonprofit, self-perpetuating society of distinguished scholars engaged in scientific and engineering research, dedicated to the furtherance of science and technology and to their use for the general welfare. Upon the authority of the charter granted to it by the Congress in 1863, the Academy has a mandate that requires it to advise the federal government on scientific and technical matters. Dr. Bruce Alberts is president of the National Academy of Sciences.

The National Academy of Engineering was established in 1964, under the charter of the National Academy of Sciences, as a parallel organization of outstanding engineers. It is autonomous in its administration and in the selection of its members, sharing with the National Academy of Sciences the responsibility for advising the federal government. The National Academy of Engineering also sponsors engineering programs aimed at meeting national needs, encourages education and research, and recognizes the superior achievements of engineers. Dr. William A. Wulf is president of the National Academy of Engineering.

The Institute of Medicine was established in 1970 by the National Academy of Sciences to secure the services of eminent members of appropriate professions in the examination of policy matters pertaining to the health of the public. The Institute acts under the responsibility given to the National Academy of Sciences by its congressional charter to be an adviser to the federal government and, upon its own initiative, to identify issues of medical care, research, and education. Dr. Kenneth I. Shine is president of the Institute of Medicine.

The National Research Council was organized by the National Academy of Sciences in 1916 to associate the broad community of science and technology with the Academy's purposes of furthering knowledge and advising the federal government. Functioning in accordance with general policies determined by the Academy, the Council has become the principal operating agency of both the National Academy of Sciences and the National Academy of Engineering in providing services to the government, the public, and the scientific and engineering communities. The Council is administered jointly by both Academies and the Institute of Medicine. Dr. Bruce M. Alberts and Dr. William A. Wulf are chairman and vice chairman, respectively, of the National Research Council.

The project was supported by the Department of the Interior Cooperative Agreement No. 1443CA000197005. Any opinions, findings, conclusions, or recommendations expressed in this publication are those of the author(s) and do not necessarily reflect the view of the organizations or agencies that provided support for this project.

Library of Congress Catalog Card Number 97-80670
International Standard Book Number 0-309-05989-5

Cover photographs: Dale R. McCullough, Berkeley, California

National Academy Press (http:/www.nap.edu)
800-624-6242
202-334-3313 (in the Washington, D.C. metropolitan area)

Printed in the United States of America

BRUCELLOSIS
IN THE GREATER YELLOWSTONE AREA

NORMAN F. CHEVILLE, Principal Investigator, Iowa State University, Ames, Iowa

DALE R. MCCULLOUGH, Principal Investigator, University of California, Berkeley, California

LEE R. PAULSON, Project Director

NORMAN GROSSBLATT, Editor

KATHRINE IVERSON, Project Assistant/Information Specialist

STEPHANIE PARKER, Project Assistant

NORMAN F. CHEVILLE is Chair of the Department of Veterinary Pathology at Iowa State University. He received the DVM degree from Iowa State University (1959) and MS (1963) and PhD (1964) from the University of Wisconsin. In 1968, he served a sabbatical year at the National Institute for Medical Research, London, studying under Anthony Allison. The honorary degree *Doctor Honoris Causa* was conferred by the University of Liège in 1986 for outstanding work in veterinary pathology. Dr. Cheville began his career at Army Biological Laboratory, Fort Detrick, Md., in the Veterinary Corps of the U.S. Army, 1959-61. After 3 years as research associate at the University of Wisconsin under Dr. Carl Olsen, he moved to the National Animal Disease Center as Chief of Pathology Research, 1964-89, and later as chief of the Brucellosis Research Unit, 1989-1995, during which he led the team that developed a new vaccine for bovine brucellosis. In 1995, he was appointed chair of the Department of Veterinary Pathology at Iowa State. Dr. Cheville has been Secretary-Treasurer and President of the American College of Veterinary Pathologists, President of the Conference of Research Workers in Animal Disease, and Editor of *Veterinary Pathology*. His honors include Outstanding Achievement Award, USDA, 1991; Distinguished Scientist Award, ARS, USDA, 1990; Alumni Merit Award, Iowa State University for "outstanding contributions to human welfare and professional accomplishment" in 1978. He has published more than 200 papers and 7 books.

DALE R. MCCULLOUGH is Professor of Wildlife Biology in the Ecosystem Sciences Division of the Department of Environmental Science, Policy, and Management and Resource Conservation in the Museum of Vertebrate Zoology at the University of California, Berkeley, where he holds the A. Starker Leopold endowed chair. He received his BS in wildlife management from South Dakota State University (1957), MS in wildlife management from Oregon State University (1960), and PhD in Zoology from University of California, Berkeley

(1966). He was a Professor of Resource Ecology in the School of Natural Resources at the University of Michigan, Ann Arbor from 1966 to 1979, and in 1980 he moved to his present position at the University of California. His research interests concern the behavior, ecology, conservation, and management of large mammals, in which he has endeavored to integrate new areas of conservation biology into traditional wildlife management approaches. He has spent sabbatical leaves in the outback of Australia studying three co-occurring species of kangaroos, and in Taiwan studying the elusive Reeves' muntjac, a small forest deer. He has served previously on three NAS/NRC committees reviewing wildlife management issues. He has published more than 100 papers and five books. Among his honors are three outstanding book of the year awards from the Wildlife Society and being named a distinguished alumnus by South Dakota State University and Oregon State University.

LEE R. PAULSON is Program Director for Resource Management in the Board on Environmental Sciences and Toxicology. She has served as project director or senior staff officer for numerous National Research Council studies, including *Setting Priorities for Land Conservation, Hazardous Materials on the Public Lands, Animals as Sentinels of Environmental Health Hazards, The Evaluation of Forensic DNA Evidence,* and *Monitoring Human Tissues for Toxic Substances.*

Preface and Acknowledgments

Brucellosis in the Greater Yellowstone Area (GYA) has been the subject of much debate and national attention. In the conduct of this study, we had the opportunity to hear many views and deeply held convictions. But at the heart of the controversy over bison management is the need for a solid scientific underpinning. To that end, we directed our efforts to identifying current research and reviewing previous research. We made every effort to represent accurately consensus views of researchers and other experts. It is our hope that this report will provide a basis for future endeavors related to managing brucellosis in the GYA and that science can be melded with policy to resolve many of the difficult issues faced by the governmental parties involved in brucellosis management. Each entity has been faced with pressures to act in the best interests of their commercial or recreational users. We further hope that this report will provide a beginning for use of emerging technology to develop a plan appropriate to the task, and one that is in best interests of the nation.

We are deeply grateful to the many colleagues who have contributed data for the manuscript during its development. A great many people have been most generous in providing us with much data, information, and observations on brucellosis, on the Greater Yellowstone Area, and on a wide variety of animals and their behavior. We wish to acknowledge in particular Drs. Steven Olsen (USDA), Mitchell Palmer, Jack Rhyan (USDA), and Beth Williams (University of Wyoming) for contributing data, graphs, and photographs of brucellosis in bison. The perspective of Mary Meagher, who has spent a lifetime with bison, was invaluable. Tom Thorne and Terry Kreeger, of the Wyoming Game and Fish Department, whose studies of diseases of wildlife and initiative in organizing the Greater Yellowstone Interagency Brucellosis Commission, were responsible for much of our practical understanding of brucellosis in elk of the Greater Yellowstone Ecosystem. The work of the Greater Yellowstone Interagency Brucellosis Committee, particularly under the influence of Drs. Dan Huff and Bob Hillman, was crucial to our understanding of the cooperation and compromise that will be required to solve this brucellosis dilemma. Data obtained from the current bison study in the YNP by Keith Aune (Montana Department of Fish, Wildlife and

Parks), Peter Gogan (U.S. Geological Survey Biological Resources Division), Jack Rhyan, Tom Roffe (USGS, BRD), and Mark Taper (Montana State University, Bozeman) gave great insight into where we are going with brucellosis in *Bison bison*.

Additional thanks and appreciation are due to Joel Berger, University of Nevada, Reno; Mark Boyce, University of Wisconsin, Stevens Point; Wayne Brewster, Yellowstone National Park; Steve Cain, Grand Teton National Park; Franz Camenzind, Jackson Hole Conservation Alliance, Wyoming; Andrew Clark, State Veterinarian, Oregon; Ron Cole, Hart Mountain National Antelope Refuge, Oregon; Walt Cook, University of Wyoming; Lynette Corbeil, University of California San Diego; Don Davis and James Derr, Texas A & M University; Phil Elzer and Fred Enright, Louisiana State University; Darla Ewalt, USDA, APHIS; Phillip Farnes, Snowcap Hydrology, Bozeman, Montana; Elmer Finck, Emporia State University, Kansas; Robert Garrott, Montana State University, Bozeman; Eric Gese, National Wildlife Research Center, Fort Collins, Colorado; Mike Gilsdorf, USDA, APHIS; Scott Grothe, Montana State University, Bozeman; Sam Holland, State Veterinarian, South Dakota; Tom Lemke, Montana Department of Fish, Wildlife and Parks; John Linnell, Norwegian Institute for Nature Research; Paul Nicoletti, Florida State University; Richard Ockenfels, Arizona Game and Fish Department; Rolf Peterson, Michigan Technological University, Houghton; Paul Rebich, Bigsky Beefalo, Montana; David Sands, Montana State University; D. J. Schubert, The Fund for Animals, Meyer & Glitzenstein, Washington, D. C.; Steve Sheffield, Clemson University; Bruce Smith, National Elk Refuge, Wyoming; Scott Smith, Wyoming Game and Fish Department; M. Stewart, USDA, APHIS; Ken Taylor, Alaska Department of Fish and Game; John Weaver, Wildlife Conservation Society; and Randall Zarnke, Alaska Department of Game and Fish, Fairbanks.

Several persons gave their of their time and expertise to review this document, and we thank them for their critical input: Beverly Byrum, Ohio Department of Agriculture, Reynoldsburg; Franz Camenzind, Jackson Hole Conservation Alliance, Wyoming; Robert Ehlenfeldt, Wisconsin Department of Agriculture, Madison; Will Garner, Logan, Utah; Burke Healey, State of Oklahoma, Oklahoma City; Daniel Jarboe, Ft. Detrick, Maryland; R. Langford, Walter Reed Army Medical Center, Washington, D.C.; Robert Mead, Washington Department of Agriculture, Olympia; Gordon Orians, Seattle, Washington; David Pascual, Montana State University, Bozeman; Duncan Patten, Bozeman, Montana; George Seidel, Colorado State, Ft. Collins; Morton Swartz, Massachusetts General Hospital, Boston; Steve Torbit, National Wildlife Federation Rocky Mountain Natural Resources Center, Boulder; Fred Wagner, Utah State.

We are also grateful to Secretary of the Interior Bruce Babbitt for making this work possible, as well as administrators and scientists within the National Park System of the U. S. Department of Interior, and in the Animal and Plant Health Inspection Service of the U. S. Department of Agriculture.

Special thanks are owed to Margaret Jaeger and Thomas Kucera for their dedicated work in gathering information from scattered sources and preparation of the manuscript. Special thanks also are due to Kathy Iverson for arranging meetings, logistics, and travel and to Stephanie Parker, who designed and crafted the web page to keep the public informed of project progress.

And finally, it has been our good fortune to work with Lee Paulson as project director—incisive, enthusiastic, highly literate, and up-front, she brought to the project the capacity to shift rapidly through conflicting opinions to identify and focus on essential items.

Norman F. Cheville
Dale R. McCullough

CONTENTS

Brucellosis
in the Greater Yellowstone Area

EXECUTIVE SUMMARY

In the winter of 1996-1997, the Yellowstone National Park (YNP) bison (*Bison bison*) population was more than 3,400 animals. Harsh weather that winter forced record numbers of bison to leave the park in search of forage; other animals in the park starved. National attention focused on management strategies—including shooting bison—used to prevent the possible spread of brucellosis from park bison to cattle that are grazed on land adjacent to the park.

Brucellosis in the Greater Yellowstone Area (GYA) is a disease caused by *Brucella abortus*, a bacterial organism transmitted primarily by contact with products of birth or abortion or by milk. *B. abortus* probably is not native to North America but was introduced with European cattle and then transmitted to wildlife; it was first detected in YNP bison in 1917 and has been present ever since. Brucellosis can be transmitted from one species to another, and concern has been expressed for many years over the potential for wildlife in the GYA to spread brucellosis to cattle that graze on land in or adjacent to the GYA and for cattle then to transmit the disease to other species, including humans.

In response to public discussion of whether brucellosis transmission by bison or elk (*Cervus elaphus*) is a threat to domestic livestock and whether vaccination or other management strategies might prove useful in controlling potential transmission, Secretary of the Interior Bruce Babbitt asked the National Academy of Sciences to undertake a 6-month study of brucellosis in the GYA. The Board on Agriculture and the Board on Environmental Studies and Toxicology began the study in May 1997 to look specifically at the following issues:

- The extent of bison infection with brucellosis in the Greater Yellowstone Area and the potential of developing a vaccine program.
- The transmission of *B. abortus* among cattle, bison, elk, and other wildlife species.
- The relationship, if any, between bison population dynamics and brucellosis.
- The ability of serology testing to estimate true infectiousness.
- The efficacy and safety of existing vaccines for target and nontarget species and the need for new (including bison-specific) vaccines.
- The nature and likely successes or limitations of a wild animal vaccination program.
- Key factors in reducing risk of transmission from wildlife to cattle and among cattle.

Some claim that the possibility that bison or other wildlife transmit brucellosis to cattle is remote and that no management strategies are needed. Others claim that any risk of transmission is unacceptable for public health and economic reasons, and brucellosis must be eradicated from the wild. This study assesses the current state of knowledge about brucellosis infection and transmission, makes recommendations for further research, and examines the implications of various management options.

CHARACTERIZATION OF BRUCELLOSIS INFECTION

Brucellosis can be transmitted among species; in humans, it is usually characterized by a fluctuating body temperature. Although rarely fatal, human brucellosis is recurrent and debilitating. The success of treating individuals varies widely, and lifelong infection is not unusual. Human brucellosis is not a widespread health threat today in North America because of efforts to eradicate brucellosis in cattle and the use of sanitary procedures (such as pasteurization) in milk processing; human infection that does occur today generally is among people who handle infected tissues, such as veterinary workers and hunters. The hallmark sign of brucellosis in cattle, bison, and elk is abortion or birth of nonviable calves.

Because of its potential to be transmitted to humans, brucellosis is one of the most regulated diseases of cattle in the United States. Cattle shipped interstate are tested routinely only for brucellosis and tuberculosis, although other diseases cause markedly more morbidity and mortality. The U.S. Department of Agriculture (USDA) established the national brucellosis eradication effort in 1934 to address public health concerns and the economic consequences to the cattle industry resulting from infected herds; that effort implemented the standards for testing, quarantine, and elimination that remain in place today. Since 1934, an estimated $3.5 billion in federal, state, and private funds has been spent on brucellosis eradication in domestic livestock. The present National Brucellosis Program is run by the USDA Animal and Plant Health Inspection Service (APHIS), which has a goal of eradicating brucellosis from U.S. cattle and captive bison herds by 1998. Only 12 cattle herds were infected at the time of this report.

As part of its efforts to eradicate brucellosis, APHIS certifies states as brucellosis-free, class A, class B, or class C, depending on the rate of infection in all cattle herds in a state. No states carry class B or C status today—an indication of the success of eradication strategies. A state's classification is important because if *B. abortus* is detected, numerous costs are incurred, such as those related to testing procedures, but perhaps the most important costs are those associated with the refusal of other states to accept a state's cattle because of the perception that *B. abortus* might be present. Many states prohibit importation of unvaccinated breeding cattle.

By authorizing USDA to regulate brucellosis transmission in cattle, the federal government has demonstrated concern that although a low risk, brucellosis poses a potentially great-loss situation in terms of potential economic consequences and possible human health effects. This report was written with that federal recognition in mind.

DETECTION OF BRUCELLOSIS

When present, *B. abortus* usually is found in the organs and tissues of the reproductive system and mammary gland, associated lymph nodes, and lymph nodes of the head and neck. Bison with non-reproductive-tract infection generally do not pose a risk of transmission to elk or cattle, although there are exceptions. But it is unlikely that large numbers of animals in a herd would be infected in lymphoid tissues without also being infected in the reproductive system or mammary gland. The two most likely events during which transmission could occur are abortion or birth.

Animals are tested for brucellosis using serologic tests (blood tests to detect that antibodies are present as a result of an infection) and bacterial cultures (where bacteria from tissue samples are grown under laboratory conditions). Both methods have flaws. A serologic result can be a good indicator of infection, but because it detects antibodies, not living bacteria, it is indirect evidence of infection or vaccination. Thus, a seropositive animal might not be infectious. Some 30-40% of bison in YNP have positive blood tests for antibodies (are "seropositive") for *B. abortus*; in the Jackson bison herd, 77% of the animals sampled are seropositive. About 1-2% of elk that do not frequent winter feeding grounds are seropositive, but in some feeding grounds, the rate is much higher—about 37%—because dense concentrations of elk create conditions favorable to disease transmission.

An animal might be infected but test seronegative in several situations, such as when antibodies have not yet developed because the test is taken in early stages of disease incubation, when a test is not sensitive enough to detect low levels of antibodies, or when the test itself is defective. False-positive tests also occur.

Finding: Seronegative results do not necessarily establish the absence of infection, because some seronegative animals in chronically infected herds are carrying live B. abortus.

Bacterial culture is the definitive test of infection, but in chronic infections, such as those present in the YNP bison herd, few bacteria might be present in an animal. That makes accurate culture difficult—obtaining the correct tissue and the correct sample size can be problematic. Therefore, although bacterial culture does not yield false-positive results, it does give false-negative results.

Although high serologic responses correlate well with bacterial cultures in bison, the relationship between serologic tests and bacterial culture is difficult to ascertain, because quantitative assessments to examine the relationships have not been done. A substantial part of the differences in GYA bison between the high percent of seropositivity and the much lower percent of positive bacterial results most likely is due to culture or sampling techniques. Multiple serologic tests and bacterial cultures on the same animals over time are the most reliable method to determine infection in live animals.

Recommendation: Because of testing insufficiencies, seropositive bison should be assumed for management purposes to be carrying live *B. abortus*.

RISK OF TRANSMISSION

Much of what we know about brucellosis in the GYA has been extrapolated from research conducted on cattle. Almost no controlled research has been done concerning transmission between wildlife species and cattle. YNP and Grand Teton National Park (GTNP) bison populations are chronically infected with *B. abortus,* as are elk populations, but the true prevalence of brucellosis in GYA bison and elk is unknown. The risk of transmission is determined by the number of abortions that occur, the presence and survival of *B. abortus* in aborted tissues, and the exposure of a susceptible host. The number of abortions or fetal deaths in bison since brucellosis first was detected in the GYA in 1917 is unknown, but in the past decade, two cases have been documented. Cattle, bison, and elk are susceptible to the same strain of *B. abortus,* and transmission between species has been demonstrated experimentally. Epidemiologic evidence, particularly that from GTNP and the National Elk Refuge (NER) points to transmission between free-roaming bison and elk and cattle as well.

Finding: The risk of bison or elk transmitting brucellosis to cattle is small, but it is not zero.

Transmission of *B. abortus* from elk to cattle is unlikely in a natural setting, because elk usually avoid areas used by cattle and isolate themselves for birth, but elk are capable of transmitting the bacteria to cattle. If cattle in the GYA mingled with aborting elk on the feeding grounds (which are maintained to promote herd growth for recreational hunting, to keep elk from straying where cattle are present, and to prevent damage to private hay crops) they would be at high risk for infection because of the high abortion rate among feeding-ground elk and the high concentration of animals. Elk also can transmit the bacteria to bison, and this might have occurred in the GYA. Under present conditions, even if low infection rates were attained for bison, an elk-to-bison or bison-to-elk transmission eventually would occur. Many more elk than bison are present in the GYA.

Finding: If infection rates are not substantially reduced in elk, reinfection of bison is inevitable.

Finding: B. abortus *is unlikely to be maintained in elk if the elk winter-feeding grounds were closed.*

There is no risk of *B. abortus* transmission to cattle from bison if bison do not leave YNP. Strategies such as discontinuing road grooming (packing snow on park roads, which some believe provides an energy-efficient travel route for bison) have been suggested to relieve the need for artificial control outside the park. But an expanding bison population searching for forage is the fundamental force pushing bison out of YNP, and the bison population will continue to increase over several years until a high population combined with a harsh winter reduces the population again. (In contrast, northern-herd elk are fluctuating about a dynamic equilibrium in response to the local food-resource carrying capacity, as well as winter stress conditions.)

Finding: Brucellosis is not a major factor in herd survival for elk or bison; among natural variables, winter mortality is the most important.

Finding: Bison leave YNP as a result of an increasing population and harsh winter weather, and under current management practices within the boundaries of YNP, the bison population will continue to grow.

Other species in the GYA, such as coyotes, grizzly bears, and wolves, can be infected by *B. abortus*. The transfer of infection among elk, bison, and cattle by those species is rare, although it cannot be ruled out completely. Carnivores and predators might contribute to transmission by transporting infectious materials from one site to another, but this probably is outweighed by the fact that carnivores and predators typically sanitize a site, thereby reducing the chance of transmission.

REDUCING THE RISK OF TRANSMISSION

Although the risk of *B. abortus* transmission is low, it can be reduced further with a combination of management options. The options for dealing with brucellosis range from doing nothing to attempting eradication of the disease.

• If nothing is done, bison and elk will be infected at the balance between the rate of transmission within and among species and the frequency of natural (that is, genetic) resistance to *B. abortus*.
• If a program to control brucellosis were undertaken, a variety of approaches could be exercised, some of which could be undertaken at the same time. The approaches taken would depend on short- and long-term goals. Several approaches to control and eventual eradication of brucellosis are available, including vaccination, establishment of perimeter zones, spatial and temporal separation of cattle and bison, and vaccination with herd

management (which might include testing and eliminating infected animals). Those approaches could be used individually or combined, depending on the degree of control determined to be in the best national interest. Other possibilities for control might arise, particularly as vaccine development progresses.

• A program to eradicate brucellosis entirely would need to include an extensive vaccination effort, as well as a test-and-slaughter component with simultaneous elimination of all infected bison, elk, and cattle. If brucellosis were eradicated from those species, the reservoirs of *B. abortus* in other wild species are expected to disappear on their own. Total eradication of brucellosis as a goal is more a statement of principle than a workable program at present; neither sufficient information nor technical capability is available to implement a brucellosis-eradication program in the GYA. No good vaccine or vaccine delivery mechanism is available at present—it would be impossible to vaccinate all GYA elk, and attempts to vaccinate bison (for example, by rounding them up) likely would be very intrusive.

Some measures can be taken immediately that would provide a good first step in reducing the risk of transmission from wildlife to cattle, regardless of final goals. The concept of surveillance zones might well be applicable to risk control for brucellosis in the entire GYA. This report emphasizes bison in YNP because of the importance of that problem in recent years when bison movements in hard winters forced a response by management agencies.

Recommendation: USDA and DOI should develop a plan to maintain a series of YNP perimeter zones with progressively increasing disease surveillance, vigorous monitoring, vaccination, and contact-reporting programs as one nears the park. The boundaries of the zones and management needed to maintain the zones should be determined jointly by USDA, DOI, and the states surrounding YNP. The plan should remain in place until brucellosis is eliminated from YNP.

It is important that a team of scientists be involved in this program and that results be analyzed and published in a refereed scientific journal.

Vaccination is an essential component of any program to control or eradicate brucellosis. Two vaccines—Strain 19 (S19) and Strain RB51—are used in cattle to protect against *B. abortus* infection. The vaccines do not produce complete protection in cattle, and the data available suggest that is also true for bison and elk. However, appropriate efficacy (the ability of a vaccine to produce desired effects) and safety tests have not been conducted for bison or elk; doses for commercial bison herds follow recommended doses for cattle. There are many unknowns besides correct ranges of dose for bison and elk, including appropriate routes, duration of immunity, and age and sex differences.

Recommendation: A long-term, controlled vaccination study must be conducted to assess the complete role of vaccination in brucellosis control and eradication for bison and elk.

An effective vaccination program would aid in reaching short-term disease control measures. Any program with a vaccination component would need to account for the large numbers of elk in the GYA, the high seropositivity rates in feeding-ground elk, and the potential for reinfection of bison by elk.

Recommendation: Any vaccination program for bison must be accompanied by a concomitant program for elk.

Recommendation: If the current vaccination program in elk feeding grounds is continued, it should include collection of serologic and culture data and appropriate epidemiologic analysis.

A coordinated, phased plan could be developed for research on the vaccination of bison, with phases that begin in sequence but could occur simultaneously. Such a plan might include collection and analyses of data from commercial bison herd vaccination programs that are under way, expansion of current experimental research on characterization of candidate vaccines in bison, and development of a field vaccination study of bison that are inside the GYA, but outside YNP.

The steps beyond a vaccination program are unclear. Whether a test-and-slaughter program is needed will depend on whether eradication is a feasible and desirable end point; further research on transmission and efficacious vaccines will be needed. The outcome of maintaining perimeter zones also will be important in determining whether eradication of brucellosis in the GYA is desirable.

An adaptive management approach that had research designed to provide data to reduce areas of current uncertainty should eventually give a more realistic assessment of the feasibility of eradication of *B. abortus* in the GYA. In adaptive management, management and research are combined so that projects are specifically designed to reveal causal relationships between interventions and outcomes, that is, to maximize learning.

Recommendation: A brucellosis program for wildlife in the GYA should be approached in an adaptive management framework.

It might prove impossible for various reasons to eliminate brucellosis from bison and elk in the GYA, so the best that could be achieved would be risk control. Bison might continue to require artificial control (such as shooting bison that leave the park), either at current or redrawn lines. Nevertheless, a cooperative arrangement to pursue systematically a pragmatic program is the best route to the highest result that can be achieved.

Recommendation: Clear short-term strategies to arrive at long-term goals must be defined and agreed upon by the federal and state entities that are involved in GYA management.

Current research and funding cannot be relied upon to sustain any long-term program effectively. As is evident from the science reviewed for this report, studies have been characterized by stop-and-go funding and elusive goals. Sample sizes have been inadequate and studies have been of insufficient duration.

Recommendation: Research priorities with sufficient funding need to be determined cooperatively and with the support of the secretaries of the U.S. Department of the Interior and U.S. Department of Agriculture.

If public opinion and political directions are aligned to a common goal, and if long-term commitments can be made by the federal departments and agencies involved, it is likely that brucellosis can be eliminated from YNP without loss of large numbers of bison or loss of genetic diversity. To be successful, society and government must support, over the long term, studies that define the ecology of the GYA, develop new vaccine technologies and delivery mechanisms for bison and elk, and produce diagnostic reagents with greater sensitivity and specificity.

Other factors will affect efforts to control or eradicate brucellosis in the GYA. They are as varied as weather, environmental change, and funding for research and management in our parks. As an added variable, future shifts in public opinion could determine the fate of any eradication effort—opinion not only on how we view bison and elk, but on the acceptability of having brucellosis in the park.

BRUCELLOSIS IN THE GREATER YELLOWSTONE AREA

INTRODUCTION

At the onset of the harsh winter of 1996-1997 in the Greater Yellowstone Area (GYA)[1], the YNP bison population was more than 3,400. Record numbers of bison (*Bison bison*) left the park in search of forage, and others starved. As bison crossed into private lands and lands managed by federal agencies other than the National Park Service, national attention focused once again on management strategies—including shooting bison—used to prevent the potential spread of brucellosis to cattle that are grazed on land adjacent to the park.

Brucellosis in the GYA is a disease caused by *Brucella abortus* biovar 1, a bacterial organism transmitted primarily by contact with products of birth or abortion or by milk. In response to public discussion of whether brucellosis transmission by bison or elk (*Cervus elaphus*) is a threat to domestic livestock and whether vaccination or other management strategies might prove useful in controlling potential transmission, Secretary of the Interior Bruce Babbitt asked the National Academy of Sciences to undertake a 6-month study of brucellosis in the GYA. The Board on Agriculture and the Board on Environmental Studies and Toxicology began the study in May 1997. The study specifically addressed

---

[1]The GYA includes Yellowstone National Park (YNP), Grand Teton National Park (GTNP), and the surrounding areas in Montana, Wyoming, and Idaho (see Figures 1 and 2).

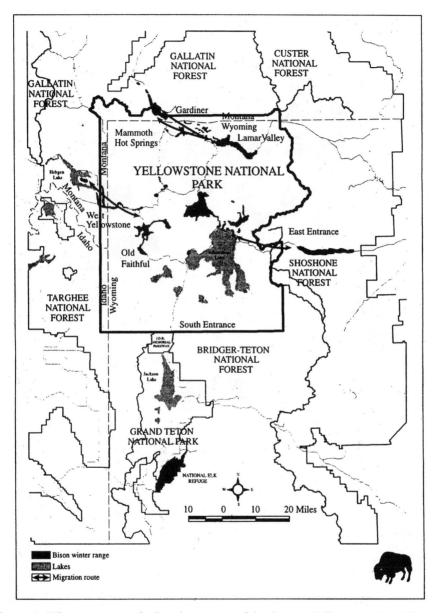

FIGURE 1. Winter ranges and migration routes of the Greater Yellowstone Area bison herds. Source: GAO 1997.

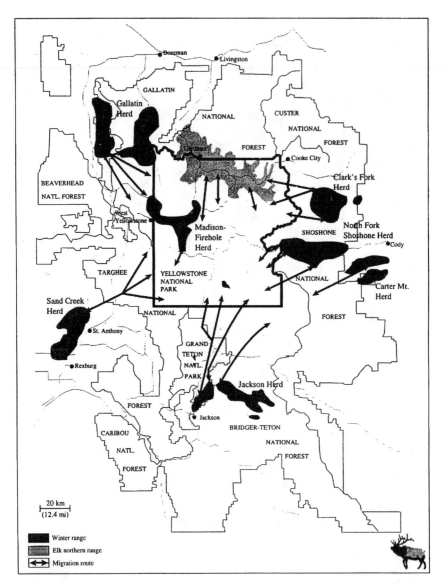

FIGURE 2. Winter ranges and migration routes of the Greater Yellowstone Area elk herds. Source: GAO 1997.

- The extent of bison infected with brucellosis in the Greater Yellowstone Area and the potential of developing a vaccine program.
- The transmission of *B. abortus* among cattle, bison, elk, and other wildlife species.
- The relationship, if any, between bison population dynamics and brucellosis.
- The ability of serologic testing to estimate true infectiousness.
- The efficacy and safety of existing vaccines for target and nontarget species and the need for new (including bison-specific) vaccines.
- The nature and likely successes or limitations of a wild animal vaccination program.
- Key factors in reducing risk of transmission from wildlife to cattle and among cattle.

BACKGROUND

Brucellosis is a zoonotic disease; in humans, it is manifested as a febrile, systemic disease, often characterized by an undulating body temperature. Although rarely fatal, human brucellosis is debilitating, and success of treating individuals varies widely. Lifelong infection is not unusual. The hallmark sign in cattle, bison, and elk is abortion or birth of nonviable calves.

YNP bison have tested positive for infection since brucellosis first was detected by Mohler in 1917. Today, some 30-40% of bison in YNP test seropositive for *B. abortus*; 1-2% of non-feeding-ground elk are seropositive. Elk at the feeding grounds have a much higher rate—about 37%—because dense concentrations of animals create conditions favorable to disease transmission.

Because of its potential to be transmitted to humans, brucellosis is one of the most regulated diseases of cattle in the United States. Cattle shipped interstate are tested routinely only for brucellosis and tuberculosis, although other diseases cause markedly more morbidity and mortality. Human brucellosis is uncommon today in North America because of efforts to eradicate brucellosis in cattle and the use of sanitary procedures (such as pasteurization) in milk processing, but it was a public-health concern in 1934. That year, the U.S. Department of Agriculture (USDA) established a national brucellosis eradication effort—which implemented standards for testing, quarantine, and elimination—that remains in place today. Since 1934, an estimated $3.5 billion in federal, state, and private funds has been spent on brucellosis eradication in domestic livestock. The National Brucellosis Program is run by the USDA Animal and Plant Health Inspection Service (APHIS), which has a goal of eradication of brucellosis from cattle and captive bison herds in the United States by 1998. Only 12 cattle herds were infected at the time of this report.

As part of its efforts to eradicate brucellosis, APHIS certifies states as brucellosis-free, class A, class B, or class C, depending on the rate of infection in all cattle herds in a state. Cattle herds in brucellosis-free states have unrestricted interstate movement. Herds in class A states have an infection rate of no more than 0.25%, and cattle must be tested for *B. abortus* before export. Class B indicates an infection rate of no more than 1.5%, and cattle must be tested before and after interstate shipment. Class C designates an infection rate of more than 1.5%, and herds must be tested twice before and once after export; no states hold class B or C status at present, which is an indication of the success of eradication strategies. A state's classification is important if *B. abortus* is detected because of numerous costs incurred, such as those related to testing procedures, but perhaps the most important costs are those associated with the refusal of other states to accept a state's cattle because of the perception that *B. abortus* might be present. Many states prohibit importation of unvaccinated breeding cattle.

STRUCTURE OF THIS STUDY

The social and political issues underlying this study are thorny and controversial. Some claim that the possibility that bison or other wildlife transmit brucellosis to cattle is remote and that no management strategies are needed. Others claim that any risk of transmission is unacceptable and that brucellosis must be eradicated from the wild. This study looks at the scientific bases behind brucellosis research and related issues in wildlife biology and makes recommendations based on current scientific knowledge.

By authorizing USDA to regulate brucellosis transmission in cattle, the federal government has demonstrated concern that brucellosis poses a low-risk, great-loss situation in terms of potential economic consequences and possible human health effects. This report was written with that in mind. The authors also are aware that the National Park Service must consider factors that are beyond the scope of this study but that might affect the ultimate management of brucellosis in the GYA, such as environmental safety of vaccines.

The principal investigators for this study, Norman Cheville and Dale McCullough, were chosen because of their expertise in *B. abortus* and in wildlife (particularly ungulate) biology, respectively. Data collection for this study had many facets, including a review of the scientific literature. A questionnaire (see Appendix A) was sent to interested parties and experts to obtain current scientific information. Open meetings were held on July 24-25, 1997, in Bozeman, Montana, and on August 4, 1997, in Jackson Hole, Wyoming, to hear scientific presentations on current research and to provide a forum at which public opinion could be expressed (see Appendix B). Experts were contacted throughout the study process to aid in synthesizing current scientific thought on issues related to brucellosis. The NRC requested public comments after the prepublication copy of the study was released in December 1997. This final document reflects changes made in response to the comments received. Changes were made to clarify the authors' text but did not result in changes to the conclusions and recommendations made in the prepublication release.

The report covers four subjects: infection, transmission of *B. abortus*, vaccination, and approaches to reducing the risk of transmission.

I
THE DISEASE AND TRANSMISSION

The same species of *B. abortus* occurs in cattle, bison, elk, and sometimes other wildlife species, and transmission of *B. abortus* among cattle, bison, and elk has occurred in captivity, which documents their common susceptibility and the potential for transmission in the wild. *B. abortus* probably is not native to North America but was introduced with the importation of European breeds of domestic cattle (Meagher and Meyer 1994).

Much research has been devoted to *B. abortus* transmission in cattle *(Bos taurus)*, but considerably less has been conducted on transmission between wildlife species and cattle. Two of the most difficult issues to address concern the probability and mode of transmission of brucellosis among wildlife species and the probability of transmission among free-ranging animals and between wildlife and cattle.

Although some studies are available and others are under way, almost no controlled research has been done on those subjects. The available evidence is essentially anecdotal and inferential, and apparent transmission in wild and free-ranging populations has been interpreted by extrapolation from what is known about transmission in domestic livestock, particularly cattle.

The following sections describe the evolutionary history of bison and factors that are important to understanding transmission of *B. abortus* in free-ranging populations. Relationships between serology and culture used to determine infection are discussed, as are environmental and animal behavior factors that affect transmission of brucellosis among species.

BISON AND CATTLE

Bison and cattle are considered to belong not only to different species, but

16

different genera (*Bison* and *Bos*). Recent studies of mtDNA (Janecek et al. 1996) suggest that bison and cattle are sufficiently closely related that they should be placed in the same genus—*Bos*—but that revision has not yet been accepted by the Nomenclature Committee of the American Society of Mammalogists, the body that sanctions such changes. Bison and cattle do have anatomic and physiologic similarities and are capable of interbreeding. They also are susceptible to similar diseases.

That bison and cattle are classified as separate species is less important than the length of time since they diverged in evolutionary history. That determines the period over which effects of natural selection, genetic drift, and, in the case of cattle, artificial selection accumulate: the longer the divergence time, the greater the expected differences between bison and cattle. The traditional view, based on paleontology and morphology, placed the diversion time at about 2 million years ago (McDonald 1981), but the more-precise DNA molecular-clock approach suggests that it is substantially greater. Examination of mtDNA control-region sequences (691 base pairs) revealed divergence of 0.09 (*Bison bison* versus *Bos taurus*) and 0.093 (*Bison bison* versus *Bos indicus*). Assuming a divergence over time of 2% per million years (Brown et al. 1979), those values imply an approximate time since divergence of 4.5 million years (J. Derr, Texas A&M University, pers. commun., 1997). That interval encompasses the evolutionary speciation of most of the currently recognized ungulate species (see, e.g., Georgiadis et al. 1990; Cronin 1991). Thus, although bison and cattle share many genes because of their common ancestry, each has been isolated for a long period during which independent mutation and selection could result in differences in physiology that equal or exceed those in morphology.

There is also differentiation below the species level, and two subspecies are ordinarily recognized (McDonald 1981): plains bison (*Bison bison bison*) and mountain or wood bison (*B. b. athabaska*). This distinction based on morphologic characters is supported by modern DNA analysis. Studies of bison at Elk Island National Park, Alberta, show that wood bison and plains bison are genetically distinct populations, based on genomic DNA restriction fragment length polymorphisms (Bork et al. 1991), and they are estimated to have diverged from a common stock around 5,000 years ago (Wilson 1969).

B. ABORTUS INFECTION AND TRANSMISSION

Much has been made of the difference in disease between bison and cattle. Certainly, *B. abortus* induces disease in bison and elk that differs from the disease in cattle. Biologic differences in antibody production, T-lymphocyte action, antibacterial proteins in normal serum, steroid-hormone synthesis, and genes for macrophage cytokines, integrins, and susceptibility genes in bison and elk all are different from those in cattle. However, the differences have not been shown unequivocally to underlie a major difference in pathogenesis that should shift our view of the pathobiology of brucellosis in the species.

Animal species vary in their clinical and tissue responses to *B. abortus*. Most bovids do not suffer marked fever, anorexia, or other signs of disease when infected. Disease is manifest only at the cellular or tissue level, such as replication of bacteria in lymphoid tissue with chronic inflammatory lesions referred to as granulomatous inflammation. The only external clinical signs might be slight swelling of lymph nodes that drain the site of infection. However, the pregnant female typically develops placental infection and can lose the fetus or experience premature labor. Even during abortion, the pain and other symptoms appear to be no greater than those encountered in normal parturition. In contrast with bovids, humans and other primates have a high, transient, fluctuating febrile response, the basis of the term "undulant fever." Lipopolysaccharide molecules on the surfaces of *B. abortus* underlie many clinical signs that occur in disease and stimulate the precipitating antibodies that are the basis of most serologic diagnostic tests.

Several major factors characterize the evolving disease in wild mammals. First, brucellosis in YNP bison is a chronic, active process. Chronicity implies a substantial degree of immunity in many individuals that blunts the course of disease and allows calves to clear *B. abortus* from their bodies before they reach sexual maturity. Second, many subpopulations of animals are present in the YNP. At one extreme are the immune calves that inherit immunity from their mothers; at the other extreme are the immunologically naive female calves—calves that, when later infected as young heifers, will carry *B. abortus* into their first pregnancy and abort a highly infected placenta. It is the latter group, typically small in a chronically infected herd, that will sustain brucellosis in the herd.

The general principles of etiologic-agent transmission in intracellular infections (Miller et al. 1994) surely apply to brucellosis. In bison herds of high endemicity, the disease occurs most often in the young; in herds of low endemicity, the disease affects all ages. The patterns of *B. abortus* pathology depend on the degree of endemicity. Where endemicity is high, even though females are still infected after the age of 5 yr, the frequencies of disease and abortion are greatly reduced. Similar immune protection from disease in older animals is never reached in herds where there is low exposure to *B. abortus*.

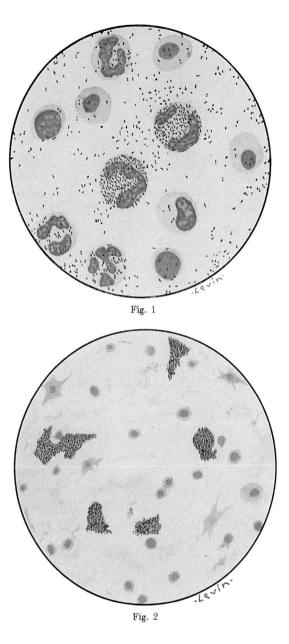

Fig. 1

Fig. 2

PLATE 1. (upper) Drawing of the microscopic appearance of *Brucella abortus* in placental exudates and placental membrane surfaces (lower) of a cow that had aborted. Bacteria are in and around leukocytes in pus. Source: Poppe 1929.

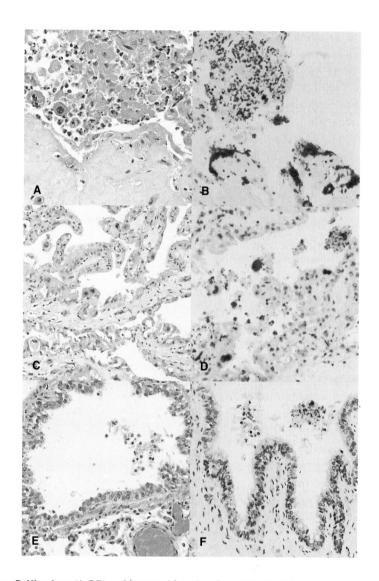

PLATE 2. Histology (A,C,D) and immunohistochemistry (B,D,F) of tissues from the uterus and retained placenta from an aborting bison cow found in March 1995 adjacent to the YNP. A. Necrosis and exudate, placenta. B. Strong labeling of bacteria in trophoblastic epithelial cells. C. Non-necrotic ares of the placenta. D. Lesser amounts of bacteria in non-necrotic areas of the placenta. E. Lung: hyperemia with necrotic debris in a large bronchiole (center). F. Bacterial debris in bronchiolar lumen are stained. Source: Courtesy J. Ryhan.

Pregnant females, even if previously immune, will have a substantially higher risk for developing pathologic lesions and abortion.

B. abortus is most likely to be found in the organs and tissues of the reproductive system and mammary gland, associated lymph nodes, and lymph nodes of the head and neck. If not present in those sites, B. abortus in bison is largely confined to lymph nodes, spleen, and other lymphoid organs, sites with fewer bacteria per gram of tissue than would be the case in the reproductive tract. Bison with non-reproductive-tract infection rarely develop bacteremia, nor do they shed bacteria in saliva, urine, or other body secretions. Thus, even when such animals are in direct contact, bison-to-cattle transmission and even bison-to-bison transmission will occur rarely, if ever. Bison with non-reproductive-tract infection do not generally pose a risk of transmission to elk or cattle (E. Williams, U. Wyom., pers. commun., 1997).

That characterization of low risk is based on the small amount of data on bison, considerable evidence of transmission of B. abortus in cattle and other ruminant species, and knowledge of other Brucella species in mammals. Although more data on bison are needed, it is imprudent to assume a high risk of transmission in bison that are not infected in the reproductive tract or mammary gland.

The overriding importance of pregnancy and the reproductive system in the life cycle of B. abortus must be understood. Brucellosis in bovids and other ruminant animals is maintained in nature by the capacity of Brucella species to grow in the female reproductive tract, particularly in membranes and fluids that surround the developing fetus (see plates 1 and 2). The natural spread of brucellosis in ruminants is highly associated with abortion and the birthing process. During the perinatal period, several events come together to define the prevalence and survival of the disease in nature.

During late pregnancy, bacteria efficiently replicate in specific epithelial cells of the developing fetal trophoblast (Figure I-3). The association of two factors drives the transmission of B. abortus and ensures perpetuation of the disease:

• Massive numbers of B. abortus in the placental fluids and genital exudates from the aborting female.
• The strong attractant effect of expelled fetal membranes.

At the time of abortion or birth of an infected calf, transmission of B. abortus to other animals occurs through contact of oropharyngeal tissues of a susceptible host with contaminated materials, usually by touching, licking, or ingestion of placental membranes and fluids. In particular, male and

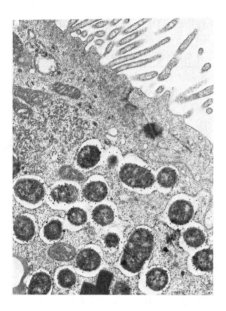

FIGURE I-3. Electron micrograph of *B. abortus*-infected chorioallantoic cytotrophoblast. Inside these cells, bacterial growth occurs in a distinct cellular compartment, the rough endoplasmic reticulum (RER), an organelle designed for highly efficient protein production. The production of peptides and glycosylation reactions in this organelle may be required for the massive growth of *B. abortus* that produces the billions of bacteria that are required for transmission to be sustained in nature (Anderson et al. 1986). Bacteria stained with an immunogold procedure to identify the genus *Brucella*.

female 2-year-olds in the nursery herd sniff and lick the expelled placenta and its fluids. Once in contact with mucous membranes of the eye, nose, or oral cavity, bacteria are taken into the body through several portals of entry, including tonsils, oronasal lymphoid tissues, and tear ducts. After the bacteria have passed the epithelial barriers of the oropharynx, they drain from the sites of initial infection into the lymphatic system and bloodstream. During travel through the lymph nodes and lymphatics and blood vessels, bacteria are taken into white blood cells, where they persist. Survival and dissemination in circulating white blood cells lead to infection of other tissues. In sexually mature animals, *brucellae* and brucella-infected cells have a strong tropism for the genital tissues and mammary glands—tissues that are needed for efficient replication.

It is unlikely that large numbers of bison in a herd would be infected in lymphoid tissues but not in the uterus or mammary glands.

However, five important cases that increase the risk of transmission when reproductive tract or mammary gland infection is absent must be noted: persistence in pregnancy, chronic infection, transmission to scavengers and predators, shedding in mammary glands and milk, and shedding in feces.

Persistence in Pregnancy

The most important exception to the rule of low-risk infections is that female bison infected only in non-reproductive tissues constitute a population of animals that can change to a high-risk population during pregnancy. The small numbers of *B. abortus* in lymphoid organs are stimulated to replicate during pregnancy and to infect the reproductive tract. Here they replicate to high numbers, and the affected female bison has a very high likelihood of transmitting *B. abortus*. The percentage of chronically infected female bison that will develop infections of the placenta and fetus during pregnancy is unknown, but the percentage cannot be assumed to be insignificant and does represent a major way in which brucellosis might continue in a population of bison.

Periodic Bacteremia in Chronic Infection

Chronically infected female cattle periodically and transiently become bacteremic and shed *B. abortus* in genital infections (Manthei et al. 1950; Lambert et al. 1960) (Figure I-4). The small numbers of positive bacterial cultures that are obtained from a chronically infected herd do not come from the same individuals; that clearly indicates that the number of infected females is always greater than the number that are shedding or bacteremic at a given time. The assumption is that the reproductive tract is infected, and the rise and fall of genital bacterial numbers are related to unknown stimuli or stress factors. However, whether a female bison must be infected in reproductive tissues for transient bacteremia to occur is unknown. If external stimuli can activate *B. abortus* growth in non-reproductive tissue that leads to asymptomatic bacteremia and genital infection in bison, the risk of transmitting *B. abortus* increases markedly for females with non-reproductive-tract infections.

Transmission to Scavengers and Predators

The risk of transmission of *B. abortus* in non-reproductive tissues from bison or elk to predators is high. Scavengers that eat infected tissues of dead bison can become infected and then, in turn, shed *B. abortus* (see Part II). Hunters and butchers are also at some risk from these animals. Predators or scavenging animals are particularly at risk of infection if they have contact with or ingest a heavily infected non-reproductive organ, such as a *B. abortus*-infected granuloma, spleen, or lymph nodes. As hosts, predators and scavengers tend

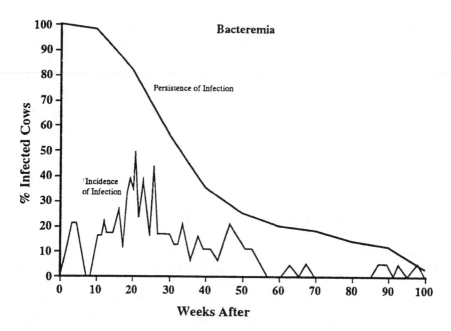

FIGURE I-4. Incidence and persistence of bacteremia in 18 experimentally infected cows (*Bos taurus*) over a 2-year period. Cows were exposed via the conjunctiva at doses of 7.4 x 10^5 *B. abortus* strain 2308 (Manthei and Carter 1950). Each sample consisted of 100 ml. blood. Incidence is the percent of cows that had positive blood samples. Persistence is the percent of cows that were known to be infected at a particular point (note that not all were bacteremic).

to be only short-term shedders of *B. abortus* and are not important in the epidemiology of brucellosis (see discussion below). The infecting doses in the tissues of non-reproductive organs are small, and the predators involved are generally viewed as "dead-end" hosts—they do not shed organisms in amounts or at sites that are likely to infect ruminant species of animals. Thus, the risk of transmission of *B. abortus* from bison or elk that are infected only in the lymphoid system to an animal population that can maintain the disease in its own species and transmit it to other species is very low. Transmission, if it occurs, might be insufficient to maintain brucellosis in any animal population.

Although fetal membranes and fluids can contain the massive amounts of bacteria that effectively promote transmission, tissues of aborted bison fetuses are also highly infectious for predators. Unlike tissues of infected adults, the lungs and gastric contents of bison fetuses typically contain the most bacteria, probably from fluids moved into the alimentary and respiratory tract during intrauterine life. In natural infections of cattle, meninges and choroid plexus of the fetus are tropic for bacteria (Hong et al. 1991), although the incidence of brucellar meningitis is not known and the mechanisms of infection are not understood.

Shedding in Mammary Glands and Milk

An important exception to the low risk of transmission concerns the mammary gland, which is not anatomically part of the reproductive tract. *B. abortus* in the mammary gland of lactating females replicates to high numbers in mammary tissue and in the lymphoid tissues of the regional lymph node—the supramammary lymph node—that drains the mammary gland. Although bacterial numbers are lower than in the infected placenta, they are typically high enough to present a serious risk of transmission. Any susceptible calf that suckles can become infected (vertical transmission[1]) even though horizontal transmission in this scenario would be highly unlikely (D. Davis, Texas A&M, pers. commun., 1997). In suckling bison calves, the relationship between oral infection via bacteria in milk and immunity from colostral antibodies is not known. Whether bacterial antigens and antibodies in intramammary milk interact in some unknown way to create complexes that increase either infection or immunity is an important pathogenesis issue that has not been well researched. Milk is an important route of transmission; further information is needed to understand the role it has in transmission.

In most lactating ruminant species, *B. abortus* is shed frequently in the milk when bacteria are localized in the mammary gland or supramammary lymph node (Morgan and McDiarmid 1960; Duffield et al. 1984). Bacterial numbers in milk are increased by a failure to suckle; the ensuing milk stasis leads to a marked increase in numbers of *B. abortus* in the mammary gland of goats and cattle (Meador et al. 1989) (Figure I-5). The association of milk stasis with a marked increase in *B. abortus* in mammary tissue and milk is an important possibility in bison, but needs to be confirmed (Rhyan et al. 1997).

---

[1]*Vertical transmission* is across generations; *horizontal transmission* is within generations.

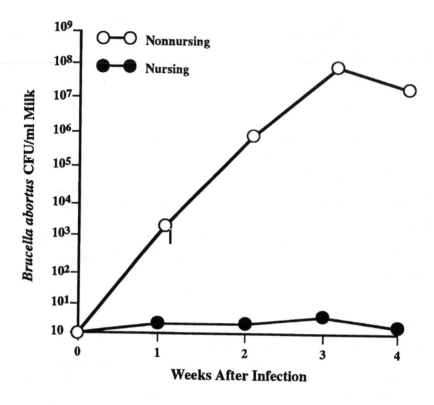

FIGURE I-5. Kinetics of bacterial numbers in mammary gland. Mean colony-forming units (CFU) of *B. abortus* isolated from milk of infected, nursing and nonnursing goats (data from Meador et al. 1989).

Shedding in Feces

Fecal shedding of *B. abortus* has been documented in bison. It is believed to occur only transiently in females that have ingested highly infected placenta and fetal fluids. In cattle, fecal shedding in the period after abortion has been reported (Fitch et al. 1932) and is thought to occur only in females consuming placenta during abortion. Because copraphagous animals could be infected by eating bison feces, fecal excretion that results from infection of intestinal tissue—particularly the lymphoid tissue of the ileum and colon—cannot be ruled out.

INFECTION IN MALES

Bison bulls develop orchitis, epididymitis, and seminal vesiculitis when infected with *B. abortus* (Tunnicliff and Marsh 1935; Williams et al. 1993; Rhyan et al. 1998); this suggests that excretion of bacteria from the testes, epididymides, and seminal vesicles into semen and urine could pose a risk of venereal transmission. However, shedding of *B. abortus* in semen of bison is reportedly rare (M. Stewart, APHIS, pers. commun., 1997), and the presence of *B. abortus* in bison semen and urine is not adequately documented.

In a current surveillance study of bison that were killed while leaving YNP in the winter of 1997, *B. abortus* biovar 1 has been isolated from testes, epididymides, and seminal vesicles of 50% of culture-positive animals; in this survey, 57% of the YNP male bison are culture positive for *B. abortus* to date. *B. abortus* biovar 1 also has been isolated from testicular abscesses in two YNP bison (D. Ewalt, Nat. Vet. Serv. Lab pers. commun., 1997).

In some captive herds, seropositive male bison have a high incidence of infection, that is, have culture-positive tissue (Rhyan et al. 1997). In studies on a captive herd of approximately 3,500 bison in central South Dakota, tissue and blood specimens were obtained from seven 3- and 4-year-old male bison that had been seronegative but had recently become seropositive for brucellosis. Six bulls had high serum titers and lesions in genital tissues, and *B. abortus* was cultured from all six. One bull had no observable tissue lesions and a low serum antibody titer, and *B. abortus* was not cultured from genital tissues.

Painful lesions in genital organs of bull bison appear to affect fertility and libido; males with painful testes do not compete successfully in breeding. That, with lameness that can result from brucellar arthritis and bursitis (Tessaro 1987), reduces the ability of males to breed.

Although it can occur, venereal transmission probably does not play a large role in the maintenance of brucellosis in bison in nature. Nonetheless, potential transmission through copulation does need to be considered for bison cows and domestic cows, given that bison bulls will serve domestic cows if confined with them. Most of the original hybrids between bison and cattle (beefalo) were the result of natural breeding rather than artificial insemination, the usual practice now (P. Rebich, Bigsky Beefalo, pers. commun. 1997). Rebich believes that cross breeding is unlikely to happen on the open range. Some bison bulls occasionally remain on the winter range during the summer in the northern Yellowstone area, but this can be prevented by hazing stragglers back to the summer range.

Bulls are not considered to be a likely source of infection either in the literature or in questionnaire results. Robison (1994) reported that venereal transmission by an infected bison did not occur, but the sample size was small.

No studies of transmission of *B. abortus* during copulation have been done on bison. Studies of possible transmission in cattle have been carried out over many years because at the turn of the century brucellosis was considered to be a venereal disease (Buck et al. 1919). Later work on penned brucellosis-infected bulls and clean cows has shown that transmission from bulls is usually not through copulation, but through nasal-oral contact or consumption of contaminated feed. Robison (1994) reported that venereal transmission by an infected bison bull in service of cows in a captive group failed.

Artificial insemination studies have shown that infected semen placed in the uterus often leads to transmission, whereas semen placed in the vagina or cervix usually does not (Manthei et al. 1950). Anatomy might limit the deposition of semen from bison bulls to the anterior vagina of bison cows, although that has not been established. The epithelial lining of the uterus differs from that of the vagina. Uterine epithelium is more susceptible to bacterial infection and has cellular mechanisms for bacterial uptake that are absent in vaginal epithelium. The low numbers of *B. abortus* shed in semen probably influence the likelihood of transmission (L. Corbeil, Univ. Calif., pers. commun., 1997).

Semen of some persistently infected male cattle does contain *B. abortus*, and bison semen might be a source of transmission. In cattle, *B. abortus* was isolated from 90 of 93 consecutive semen ejaculates from one bull over 5 years (Lambert et al. 1963). Despite that presence in semen, venereal transmission was not viewed as important. *B. abortus* was not transmitted in cattle during the use of an infected bull in natural service for 3 years, although *B. abortus* was present in each of 80 ejaculates collected over 18 months; the number of organisms varying from 100 to 49,500 per milliliter of semen (Manthei et al. 1950). When the same semen was experimentally inoculated directly into the uterus, infections were induced in six of six first-calf heifers and in two of six second-calf cows.

If *B. abortus* in bison follows transmission patterns of other ruminant species, venereal transmission alone would not be sufficient to sustain the persistence in the herd. The transmission of *B. abortus* by male cattle is better understood (King 1940; Manthei et al. 1950; Rankin 1965); it is capricious and not viewed as important in the disease at the herd level (although the mechanism of persistence in male genital tissues remains unclear).

As a practical matter, brucellosis commonly is eliminated from cattle herds by vaccinating cows without regard to bulls, which suggests that venereal transmission is rare. Brucellosis has been eliminated from many managed bison herds using the same protocols as are applied to cattle (i.e., vaccination and culling).

Practical experience suggests that transmission by bulls to cows by service among bison in the wild is unlikely, but more research data are needed on the role of the male in transmitting infection, particularly to determine the age, route, and persistence of *B. abortus* in genital tissues of the male and the reliability of serologic testing to detect infections. But given the spatial separation that usually results from management and the behavioral barriers to copulation between species, transmission of brucellosis between bison bulls and domestic cows in the GYA appears to be vanishingly small.

DETECTING INFECTED ANIMALS

Like tubercle bacilli and other intracellular bacteria, *Brucella* species have the capacity to survive in the face of strong host immune defenses by entering and surviving inside the cell. The capacity for a few brucellae to persist in lymphoid tissues in chronic infections and to be reactivated to grow again is closely related to their ability to survive in nature and explains many of the seemingly capricious events that occur in serologic and bacteriologic studies. The cell in the host's lymphoid tissues that perpetuates bacterial growth in chronically infected males and nonpregnant females is not known but is often referred to as the macrophage. Coupled with long-term survival in the host is the capacity to undergo striking growth, when activated by progesterone or some other hormone associated with developing pregnancy, and to move to and replicate in the developing placenta.

Serology

The presence of antibrucellar antibodies in serum of an infected animal indicates that infection is present or that infection has occurred recently. Because they detect antibodies, not living bacteria, serologic tests are indirect evidence of infection—evidence of an immune response to live *B. abortus* or to brucellar antigens of dead bacteria that have remained in the host. In brucellosis, humoral immune responses fail to eliminate or even control infection in the host but do limit the extent of infection.

Antibrucellar antibodies in serum of bison and elk are detected by a battery of tests that include the standard tube agglutination test, complement fixation test, CARD test, rivanol test, and particle concentration fluorescence immunoassay (PCFIA). All of these use brucellar lipopolysaccharides (LPS) as the test antigen, and thus detect antibodies to the perosamine residues (O side-chains) of bacterial LPS. Rough strains of *B. abortus*, such as the mutant vaccine strain RB51, are deficient in the O side-chain of LPS and do not induce antibodies that are detected in the standard serologic tests for brucellosis. None of the LPS-based serologic tests discriminate between vaccine strain 19 and field strains of *B. abortus*; they may detect differences in magnitude or persistence of brucella antibodies induced by those strains but not qualitatively different responses.

Specific serologic data from bison and elk are being analyzed to confirm the adequacy of guidelines that are used to interpret serologic reactions. The Brucellosis Scientific Advisory Subcommittee of the U.S. Animal Health Association is evaluating data on two new tests: the rapid automated presumptive (RAP) test and the fluorescent polarization assay. The subcommittee also is reviewing an APHIS-sponsored analysis of data from sera collected from cervids (M. Gilsdorf, APHIS, pers. commun., 1997). Sections of *Brucellosis Eradication: Uniform Methods and Rules* (USDA 1984) are being updated with rules that apply to commercial bison herds, interstate shipment of bison, and bison quarantine facilities.

Serologic tests of blood and milk are used to identify infected individuals and to characterize a herd. A positive serologic test in one individual is taken as indirect evidence that the herd is infected. An infected herd is one in which at least one animal has been shown to be infected; the diagnosis is based on results of several serologic tests, bacterial culture results, and information regarding herd history, clinical signs, and epidemiology. Minimal criteria for a diagnostically positive reaction to various serologic tests in cattle are provided by APHIS. Diagnostic criteria for bison and elk have been proposed but not approved (M. Gilsdorf, APHIS, pers. commun.,1997).

Serologic responses in bison might develop more slowly than in cattle or other species. Experiments have shown that although vaccinated bison challenged intraconjunctivally with a virulent *B. abortus* strain seroconvert on brucellosis-surveillance tests (Olsen et al. 1997), the antibody responses of bison to *B. abortus* challenge lagged approximately 2 to 3 wk behind that seen in cattle at the same time (Davis et al. 1990). An animal with natural resistance to *Brucella* spp. that has been infected with *B. abortus* and has cleared the bacterium generally will have a short-lived antibody response.

The False-Negative Serologic Test—Serologic tests can be falsely negative in infected bison when there is an absence of antibodies in an infected animal, when the test is insufficiently sensitive to detect low antibody titers, or when the test being used is itself defective.

Bison can lack antibodies in their serum but still be infected. In acute infection, that occurs when serum is taken during the incubation stage when antibodies have not developed. Not all actively infected animals have serologic evidence of exposure (R. Zarnke, Alaska Dept. Fish and Game, pers. commun., 1997), and that is most often the case for serologic tests taken in early stages of incubation. In chronic infection, false-negative reactions occur when bacteria are sequestered in lymphoid tissues in a state that does not induce antibody formation.

One problem of brucellosis serology is that some animals do not clear bacteria but retain them in their lymphoid tissues in very small numbers or in an inactive state that does not stimulate precipitating antibodies sufficient to react in serologic tests. Such animals often are seronegative but must be considered infected. It is clear that bison can be infected but be seronegative on all standard LPS-based serologic tests (Figure I-6). Thus, for bison-management plans, seronegative bison cannot be assumed to be free of brucellosis.

The False-Positive Serologic Test—Serologic tests can be read as positive when no antibodies resulting from exposure have occurred. A false-positive test might be due to cross-reacting antigens or to defective test procedures. Tests also can be positive in the recovery phases in young animals that have no live bacteria.

Bacterial Culture

The isolation of *B. abortus* by bacterial culture of animals tissue or body fluids is the definitive indication of infection. Failure to culture bacteria that are present in small numbers in tissue can be due to inappropriate sampling, improper storage of specimens, or failure to use sufficient amounts of tissue. Correct techniques involve use of special growth media (Alton et al. 1988). In chronic infections in which very few bacteria are present, the use of subculture techniques might be required (Jensen et al. 1995). Any failure to follow established bacteriologic techniques will diminish the capacity to isolate *B. abortus* and, in turn, in the ability to identify all infected animals in a herd.

Other pitfalls in the interpretation of bacterial culture data are related to

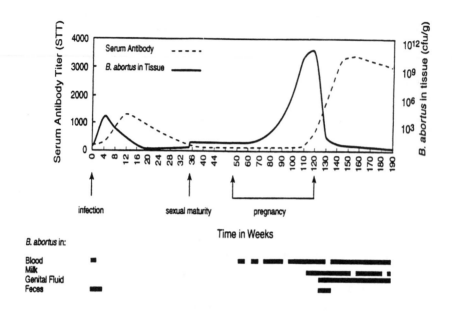

FIGURE 1-6. Hypothetical titers (based on limited data) of serum antibodies and tissue bacteria in a bison calf from birth to 190 wks of age. Early phases are characterized by bacteria in lymphoid tissues of the head and neck. There is a small peak at sexual maturity as the reproductive system is infected and a large increased in late pregnancy as the placenta, fetus, and mammary gland are infected.

the sample. Even with use of proper technique, brucellae grow very slowly; in field specimens, overgrowth of more rapidly growing pathogens and sapro phytes in the sample easily can obscure more slowly growing brucellae.

The amount of body fluid or tissue in a sample correlates directly with the detection of B. abortus. In acute infections, especially of the reproductive tract and mammary tissues, that might not be critical, because large numbers of organisms are present. But in chronic infections with few organisms per gram of tissue, large samples must be used, e.g., 100 mL of blood, entire lymph nodes, and large sections of organs. Bacteria can reside only in one small part of a lymph node, so the entire lymph node must be sampled for appropriate detection.

Tissues collected for culture are those in which B. abortus is most likely to be found: organs and tissues of the genital system and its associated lymph nodes and lymph nodes of the head and neck. The placenta, if present, should be carefully examined to note the extent of lesions in cotyledons (tufted areas of the placenta) and intercotyledonary placenta. Numbers of *B. abortus* present will be greatest in portions of the placenta that have exudates and tissue lesions. If the objective is to identify infection, then the simple determination of the presence or absence of bacteria is sufficient. Some studies, however, require precise knowledge of the extent of infection; in that case, the number of bacteria per gram of tissue should be determined, and this requires titration of tissue suspensions.

In chronically infected bison and elk, *B. abortus* is usually present in one or more lymph nodes. The specific lymph nodes to be harvested for bacterial culture should include bronchial, hepatic, internal iliac, mandibular, mesenteric, parotid, popliteal, prefemoral, retropharyngeal, superficial cervical (prescapular), and scrotal or supramammary.

The number of samples needed to detect infection in adult bison is un-known. One report from tests in adult cattle suggests that 100% detection can result from sampling of uterine caruncle and supramammary, mandibular, and medial iliac lymph nodes (Alton et al. 1988) but most laboratories do not achieve this detection rate. Young animals clear *B. abortus* quickly, and they require greater tissue sampling than adult bison.

Young adults of most species are most often culture positive because of more recent exposure and other factors. For example, *B. abortus* is stimulated to grow at sexual maturity and typically disseminates to the mammary gland, reproductive tissues, and associated lymph nodes. The growth stimulates a rise in antibody during and after sexual maturity. Thereafter, the serum antibody rises and falls according to the persistence or clearance of bacteria from tissue.

Detection of one or two *B. abortus* in lymphoid tissues of chronically infected animals remains a problem in that defects in sampling or culture technique often lead to false-negative results. A newly developed polymerase chain reaction can detect very small amounts of bacterial DNA (Bricker and Halling 1995), but it has not been established for official use.

If a highly sensitive test to detect all *B. abortus* cannot be developed, a test that will induce exacerbation of persisting bacteria might allow positive culture with current techniques—exacerbation mimics natural phenomena in brucellosis, in which unknown factors by unknown mechanisms stimulate latent *B. abortus* to replicate and spread in the host.

Correlation of Serology with Bacterial Culture

Multiple serologic and bacteriologic culture tests done over time are the only reliable method to determine infection in live animals. A culture-positive animal always has the potential of transmitting the disease, but false-negative culture results can be obtained if inappropriate tissues are selected for culture or if tissues are mishandled during collection or laboratory processing.

High serologic responses correlate well with isolation of *B. abortus* on bacterial cultures; that is, bison infected with large numbers of bacteria typically have high serologic titers. A serologic result can be a good but not infallible indicator that an animal is infected. It is unlikely that a serologic response positive for *B. abortus* will provide a strong indication of whether an individual animal is infectious (S. Olsen, USDA, pers. commun., 1997).

The percentage of animals in a herd with serum antibodies ("positive" serology) is referred to as seroprevalence. It is widely believed that seroprevalence overestimates the prevalence of *B. abortus* infection, but that fact has not been established. Current knowledge of serologic reactions to organisms of the genus *Brucella* suggests that although we know that the presence of antibodies in an animal having had brucellosis lasts beyond the point where bacteria have been cleared from the host, it is more probable that any discrepancy between a clearly positive serologic test and bacterial culture is due to culture techniques.

Tests designed for cattle have been used for years to detect seropositivity in bison, but diagnostic tests used now for cattle have not been validated in bison. Current official tests are based on LPS of *B. abortus*. Data on the serology of bison with those tests are insufficient to support dogmatic statements regarding known relationships among serology, culture-test results, and likelihood of infectiousness (Olsen et al. 1998). S. Olsen (USDA, pers. commun., 1997) stated that "as is the case in cattle, it is unlikely that a strong correlation will be found between positive responses by bison on *Brucella* serologic tests and culture-positive status."

In summary, negative serologic test results do not equate with the absence of infection. Individuals can falsely test positive or negative on serologic tests. Serologic tests are evaluated in terms of sensitivity, specificity, and predictive values instead of absolutes, and all have some degree of imperfection. It is important to recognize that discrepancies between serologic and bacteriologic data might be real or artifactual. The discrepancy in GYA bison between the high percent of seropositivity and the much lower percent of bacterial isolations is most likely due to culture techniques.

Immunity

Immunity to brucellosis could be measured in bison and elk by means other than serology. Cell-mediated immunity, as measured by lymphocyte proliferation assays from naturally infected and vaccinated cattle and bison plays a role in recovery from *B. abortus* infection (Davis et al. 1990; Olsen et al. 1997). Cutaneous delayed hypersensitivity reactions against brucellin, an extract of *B. abortus*, are used in Europe to diagnose brucellosis in cattle of seronegative herds. In the course of bovine brucellosis, cutaneous reactivity develops later than do antibody responses but persist long after serologic evidence of infection has disappeared (Bercovich et al. 1990; Cheville et al. 1994). The current recommendation is intradermal injection of 0.1 mL brucellin in the neck or tail fold and gross examination of skin lesions at 24-72 hours later. The intensity of the reaction is based on the degree of skin thickness. This test is complicated by the occurrence of false-negative reactions and by variability in evaluations; it has never been tested in bison or elk.

LIKELIHOOD OF INFECTIOUSNESS

To be infectious, an infected animal must release *B. abortus* from its body in a way that will infect another animal. To place infectiousness in appropriate context, attention must be given to its precise definition and to the definitions of *infection* and *disease*:

> **Disease**: any deviation from or interruption of the normal structure or function of any part, organ, or system of the body that is manifested by a characteristic set of symptoms and signs and whose etiology, pathology, and prognosis may be known or unknown.
> **Infection**: 1. invasion and multiplication of microorganisms in body tissues, which may be clinically inapparent or result in local cellular injury due to competitive metabolism, toxins, intracellular replication, or antigen-antibody responses. 2. an infectious disease. Cf. infestation.
> **Infectiousness**: ability to transmit a pathogenic agent from an infected individual to another susceptible individual. (Dorland's Medical Dictionary, 28th edition)

Brucellosis, particularly the chronic form in which there are few bacteria and no obvious clinical signs, is a disease. Bacteria are present in the animal, microscopic pathologic tissue changes are present, and, despite their lack of effect, immune systems are at work. Chronic forms of the disease have been called subclinical, latent, or inapparent infections.

Although controlled studies have not been done, it is unlikely that prediction of infectiousness can be based on serology alone. Clearly, bison and elk may have serologic titers to *B. abortus* yet not be infectious. It is also highly probable that those species can have serologic titers and not be infected—when the animals are in a postrecovery period in which all live bacteria have been cleared but the animals are still responding to brucellar antigens that remain in their tissue. It is dangerous to assume that large numbers of seropositive animals do not carry live *B. abortus*. The occurrence of latent carriers among cattle (heifer syndrome) is widely accepted, and experimental evidence indicates that they occur among bison and elk (Thorne and Morton 1978).

MINIMUM INFECTIOUS DOSE

The minimum infectious dose (MID) of *B. abortus* in bison—the smallest number of bacteria that can initiate infection or disease—is not known. The smallest dose that will infect 100% of cattle is reported to be 15.6×10^6 live bacteria; below that dose, the infection rate is correlated with the dose (Manthei and Carter 1950) (Table I-1). In natural conditions, infection occurs with very large doses present in placenta, placental exudates, and milk. Even in cattle, the MID has never been established unequivocally. Experimentally, the same dose will typically infect a different percentage of the animals. The definitive study in cattle used five groups of pregnant heifers (10 per group) to estimate infective dose of virulent *B. abortus* given conjunctivally. The extent of placental infection was greatest in heifers given the largest inoculum (McEwen et al. 1939).

The value of the MID is compromised by two important characteristics: the validity of data used to establish it, and lack of full knowledge of the factors that skew the MID one way or the other. For example, animals with intercurrent disease that stimulates marked antibody production or cell-mediated responses show a significantly increased immune response to virulent *Brucella* spp.: calves with severe cutaneous fungal infections, such as ringworm, develop strikingly greater antibody responses to virulent *B. abortus* and clear the bacteria from the site of infection and draining lymph node much more quickly (Cheville et al. 1993) than animals without fungal infections do.

The capacity of *B. abortus* to survive in soil and debris at varied temperature, acidity, and relative humidity has a great bearing on numbers of bacteria available for transmission. Little is known about how long bacteria survive after abortion or birth events under natural conditions in the YNP. E. Williams (Univ. Wyo., pers. comm., 1997) stated that "preliminary studies indicate prolonged survival of the bacteria in the Wyoming environment with the titer of bacteria remaining high for weeks in the late winter. . . . In spring, numbers of bacteria decrease rapidly along with accelerated decomposition of fetuses."

TABLE I- 1. Incidence of infection and abortion in normal nonvaccinated pregnant bovine heifers given different conjunctival doses of virulent *B. abortus*. Challenge strains were 544 (McEwan et al. 1939) or 2308 (Manthei and Carter 1950).

Authors	Dose, no. bacteria	Proportion infected, %	Percent aborted
McEwen et al.	1.46×10^5	45	22
	1.46×10^6	50	30
	1.46×10^7	90	90
	1.46×10^9	100	77
Manthei and Carter	3.5×10^5	78	56
	7.0×10^7	87	74
	1.5×10^7	100	100
	2.5×10^7	97	91
	7.5×10^7	100	92
	1.0×10^8	100	97

The value of MIDs for bison is questionable. Experiments to obtain MIDs for bison will be expensive; once derived, they will be limited in application, because differences in bacterial dose, route, temperature, and host factors (age, sex, stress, and disease) will cause variations that exceed the value of the data.

TRUE PREVALENCE OF *B. ABORTUS* IN GYA BISON AND ELK

The true prevalence of brucellosis in GYA bison and elk is unknown. Insufficient sampling (in regard to both number and reliability) has been done to establish reliable data. Furthermore, in previous studies, some of the methods used were imprecise and led to false-negative or false-positive results of serologic and bacteriologic tests. The data that are available suggest only that the true prevalence is not zero and might vary from 12% to 100% in bison and from 1% to 38% in elk.

Bison

Bison populations in YNP and Grand Teton National Park (GTNP) can be considered to be chronically infected with *B. abortus*. Many bison develop immune responses but do not clear the bacteria (Olsen et al. 1997). *B. abortus*-induced abortion in free-ranging bison was first proved by bacterial culture and pathology in March 1989 in the Jackson herd of about 120 bison. *B. abortus* biovar 1 was isolated from reproductive tissues, and the pathologic description of disease was clear: "endometritis was characterized by lymphoplasmacytic infiltrates in the lamina propria and neutrophils in uterine glands and within necrotic debris and exudate in the uterine lumen" (Williams et al. 1993). *B. abortus* has since been shown to cause abortion in bison from YNP (Rhyan et al. 1994). The Jackson herd is infected with *B. abortus*; seroprevalence in 35 bison sampled in 1989-90 was 77%, and *B. abortus* was cultured from 4 of 11 (36%) seropositive bison (GYIBC 1997).

In 1985, *B. abortus* was cultured from 6 of 88 bison (Clark and Kopec 1985), and in 1991-92, from 26 of 218 bison (Aune and Schladweiler 1992). Recently, *B. abortus* biovar 1 was isolated from an aborted fetus found near Old Faithful and a stillborn calf located outside the YNP; both were heavily infected with *B. abortus* (billions of organisms per gram of tissue). Data on a retained placenta from a bison shot and sampled on the north side of YNP established that *B. abortus* infects the placenta and causes abortion in bison in a manner similar to its effects on cattle. Furthermore, vegetation and soil were sampled at two bison birth sites (Lamar area samples taken May 8 and 22, 1996; North Gate area samples taken May 16, 1996); *B. abortus* remained viable in soil for 14-18 days after abortion occurred (Roffe et al. 1997).

Serologic evidence of brucellosis in YNP bison was first reported in 1917, and 40-54% of YNP bison tested have been reported as seropositive since then (GYIBC 1997). Serologic evidence indicates that as many as 60% of YNP bison (Tunnicliff and Marsh 1935, Pac and Frey 1991, Aune and Schladweiler 1992) and 77% of GTNP bison (Williams et al. 1993) contain serum antibodies against *B. abortus*. Those data do not include animals that are infected but do not have a measurable antibody response at the time of sampling.

Immune responses from natural infection might induce some degree of protection against *B. abortus,* inasmuch as substantial fetal loss or infertility has not been reported for the bison populations in the GYA (Olsen et al. 1998). Experimental studies suggest that bison are more susceptible to brucellosis than are cattle or elk (Davis et al. 1990, 1991): nearly all infected female bison aborted their first calf.

Elk

In North America, substantial brucellosis in wild elk occurs only in the GYA. Seroprevalence among adult female elk in the western Wyoming feeding-ground complex has averaged 37% since 1970.[2] During herd reductions of the 1960s, 1.7% of 6,027 elk on the northern range were brucellosis test reactors; YNP elk in the northern herd have not been tested recently. Elsewhere, seropositivity (Morton et al. 1981; Smith and Roffe 1992; Rhyan et al. 1997) was

- Six of 126 elk trapped northwest of the YNP during spring of 1988.
- Less than 0.03% of 3,833 elk tested in Colorado from 1967-1976.
- None of 170 Idaho elk.
- Three of 113 Utah elk.
- Two of 178 Wyoming elk not associated with feeding grounds.

In the GYA, *B. abortus* is unlikely to be maintained in elk in the absence of the bison reservoir and if the elk winter feeding grounds are closed. Elk have much lower seropositive rates on natural winter ranges in Wyoming in comparison with elk on feeding grounds. Over time, elk would not serve as reservoirs for brucellosis in the absence of elk feeding grounds.

Research at the Sybille Research Unit in Wyoming has shown that 50-70% of female elk that become infected with *B. abortus* lose their first calf (Thorne et al. 1979); retained placentae and infertility do not occur in elk as they do in cattle (Thorne and Herriges 1992).

Transmission from infected elk in captive herds to susceptible cattle occurred when the two species were in close contact and pregnant elk gave birth or aborted (Thorne and Herriges 1992). Such transmission was believed to be extremely unlikely in normal calving on traditional elk calving ranges

---

[2]Elk vaccinated at Jackson Hole with strain 19 vaccine have titers that cannot be differentiated in standard serologic tests from those caused by field strains; elk vaccinated with strain RB 51 do not have the confounding serum antibodies.

and probably occurs only in the close confines of elk feeding grounds. "Elk do not seem to be capable of sufficient intraspecific transmission of brucellosis to maintain the disease in the population when not concentrated on feeding grounds" (Thorne and Herriges 1992).

Serologic evidence of brucellosis has been found in blood samples from elk corralled in 18 of 23 feeding grounds with seropositive rates averaging 37% in adult females (Herriges et al. 1991). Blood samples from hunter-killed elk outside the GYA that were tested in 1970-1992 were negative (Thorne and Herriges 1992). Normal calving behavior—calving in seclusion and clearing the placenta—almost completely removes the likelihood that an elk will transmit brucellosis to another animal (see p. 45, "Bison and Elk Behavior and Transmission") (Thorne and Herriges 1992).

Infection in Other Mammals in the GYA

Canids

The assessment of risk of carnivore transmission involves three separate issues: the risk of transmission from bison or elk to the carnivore, the probability that infection will be maintained in that species, and the frequency of transmission back to bison or elk.

B. abortus has been isolated from wild carnivores in areas that contain infected bison and elk (Tessaro 1986). Those predators consume infected elk and bison meat, especially during the winter and early spring, and frozen infectious material distributed by scavengers cannot be overlooked (A. Clark, Ore. Dept. Agricult., pers. commun., 1997). Coyotes (*Canis latrans*) can be infected with *B. abortus* (Davis et al. 1979, 1988), but serologic surveys of coyotes have not revealed exposed animals in the GYA (Gese et al. 1997). No surveys of coyotes on feeding grounds have been reported, but exposure might occur. There is no evidence that coyotes are important in the epidemiology of brucellosis in the GYA, although coyotes elsewhere have been found to be infected and able to shed the organisms for a short period (Davis et al. 1979, 1988).

Dogs infected with *B. abortus* typically do not develop clinical signs of disease, although dogs with brucellosis do suffer abortion, epididymitis, and lameness associated with joint lesions (Forbes 1990). Seroconversion can occur 4-14 days after exposure, and a positive serotiter can be maintained for as long as 2 years (Kiok et al. 1978). Seronegative, culture-positive dogs have been reported. In a study of 14 dogs from farms with *B. abortus*-infected cattle,

B. *abortus* was cultured from lymph nodes draining the head, neck, lungs, and intestines, and urogenital infection with shedding was seen in one bitch (Forbes 1990). The maximal duration of infection was 539 days; that suggests that dogs have the potential to infect bison and elk. Dogs have also been experimentally infected orally with vaccine strains of *B. abortus*, and in pregnant females, the placenta was infected, although abortion did not occur (Palmer and Cheville 1997).

Moose

From 1937 to 1985, four cases of brucellosis in wild moose (*Alces alces*) were reported, each proved by isolation of *B. abortus* from multiple tissues. In three cases, the moose had been in contact with cattle that had brucellosis. Clinical signs in all four were weakness and debility. Lesions in the carcasses included swollen lymph nodes, pneumonia, and fibrinous exudates on the pleura, the pericardium, and other serous surfaces. Fibrinous exudates from the lungs and serous surfaces are characteristic of pasteurellosis; *Pasteurella multocida* was isolated from one of the moose (Corner and Connell 1958), so it is questionable whether brucellosis was the cause of death.

Despite the presence of brucellosis in moose, substantial titers of antibody against *B. abortus* have not been found in free-ranging moose in North America, even in areas where moose are in contact with infected cattle. That might be the origin of the widely held belief that infection with *B. abortus* is generally fatal in moose. All moose tested in the GYA and in Montana have been serologically negative, and that finding might have buttressed the conclusion that all infected moose die, leaving no residuum of seropositive, chronically infected moose.

To determine susceptibility, Forbes et al. (1996) experimentally inoculated four moose conjunctivally with *B. abortus* biovar 1. No acute phase of disease developed—no clinical signs, no abnormal blood values, no abnormal serum chemistry. Two moose were killed at day 70 after inoculation, one died at day 85, and one was killed at day 166. None of the moose had clinical signs of disease, except the moose that died, which had fever; clinical pathology data remained normal. *B. abortus* was isolated from several tissues, most notably lymph nodes, where bacterial counts often exceeded 4×10^4 CFU/g of tissue (Forbes et al. 1996). Although the authors suggest that "lesions seen in all moose were indicative of endotoxemia," no data presented supported that conclusion or established that any tissue change was caused by *B. abortus*.

If brucellosis is generally fatal in moose, it must be unusual. Acute death from brucellosis would be unique; infected ruminants do not develop life-threatening febrile disease. It has been widely claimed that moose die of a septicemic brucellosis with tissue responses that include peritonitis, pleuritis, and other acute inflammatory lesions. The evidence that such lesions are directly due to *B. abortus* is not convincing, and the pathogenesis of brucellosis in moose needs to be resolved.

It is not likely that moose play a major role in sustaining infections with *B. abortus* in wildlife in the GYA. Moose are also susceptible to *B. suis* (Dieterich et al. 1991); if they are in contact with swine or caribou, that bacterial species should be investigated.

Horse

Horses can be infected with several *Brucella* spp. and have been reported to be a source of *B. abortus* infection in cattle (White and Swett 1935). *B. abortus* has a predilection for joints, bursae, and tendons, and the common clinical features of equine brucellosis are suppurative spinous bursitis ("fistulous withers") and suppurative atlantal bursitis ("poll evil"). Although mares infected with *B. abortus* have been reported to abort, experimentally infected mares have produced normal, noninfected foals at full term (MacMillan et al. 1982). *B. melitensis* and *B. suis* (from feral pigs) have also been reported in horses (Cook and Kingston 1988). A 9-year-old pregnant mare that had contact with elk in a winter feeding ground 5 miles south of Jackson, Wyoming, developed fistulous withers that contained pus from which *B. abortus* was isolated. The mare later foaled, and neither the foal nor the placenta had evidence of *B. abortus*. A second horse used to pack out hunted elk also had fistulous withers and positive serologic tests; bacterial cultures did not grow *B. abortus*, but the horse had been treated with tetracycline before culture was attempted. Horses used by rangers and those brought into the GYA by park visitors for recreation are susceptible on contact with tissues or fluids of infected elk or bison.

Bear, Deer, and Other Big Game

Records from the Montana Livestock Sanitary Board laboratories that tabulate blood tests in big-game animals from 1932 to 1968 show that seroreactors were not found in antelope (*Antilocapra americana*), bighorn sheep (*Ovis canadensis*), mule deer (*Odocoileus hemionus*), moose, mountain goat (*Oreamnos americanus*), and grizzly bear (*Ursus arctos*); one seroreactive black bear (*Ursus americanus*) was found (Barmore 1968). Brucellosis recently was detected in black bear and grizzly bear in the greater Yellowstone ecosystem (K. Aune, Mont. Fish, Wildlife, and Parks, pers. commun., 1997). The extent of infection in bear is not known, but bear are unlikely to play a major role in the persistence of brucellosis in YNP (see Part II, "Transmission Among and Between Species").

Mule deer outside YNP have been shown to be seropositive, but deer in YNP have not been shown to carry *B. abortus*, and it is widely assumed that deer are not a major host for it. Brucellosis has not been detected in Montana in mule deer or white-tailed deer (*O. virginianus*).

II
TRANSMISSION AMONG
AND BETWEEN SPECIES

Much public discussion has centered on whether transmission of *B. abortus* in the wild ever can be documented. This section reviews epidemiologic evidence of transmission and associated factors, including the role of bison and elk behavior and the effects of weather on animal movement in the GYA. The National Park Service's natural-regulation policy is discussed, as is the effect of *B. abortus* on reproductive potential of bison.

BISON, ELK, AND CATTLE

Brucellosis was discovered in bison on first testing in 1917 (Mohler 1917), and it has existed since as a self-perpetuating disease in that species. Thus, transmission from cattle introduced by Europeans to at least one wild species must have occurred and then transmission from cattle or from the infected wild species to other wild species to account for the disease in cattle, bison, and elk (Honess and Winter 1956). Meagher and Meyer (1994) note that there were probably multiple transmissions to bison, and Thorne et al. (1991) note that recovery of *B. abortus* biovars 1 and 4 in Wyoming and the presence of *B. abortus* widely over the GYA suggest multiple exposures in elk as well. It seems likely, in view of the early free-range management of domestic stock in the West, that original transmission of the disease from livestock to bison and elk occurred during intermingling in the free-roaming state. However, at the beginning of the 20th century, restoration programs for bison (Garretson 1938) and elk (Murie 1951) resulted in capture, handling, and relocation of large numbers of both species, so the possibility of transmission in captivity cannot be ruled out.

Transmission of brucellosis from captive bison to cattle in North Dakota was reported by Flagg (1983). The strongest evidence of transmission be-tween free-roaming bison and elk comes from Grand Teton National Park (GTNP) and National Elk Refuge (NER) in the Jackson area of Wyoming (Williams et al. 1993, 1997). A small herd of bison was established in the wildlife park at GTNP in 1948 and in 1963 was found to be infected by *B. abortus*. All adults were removed, and calves were vaccinated. Brucellosis-free bison were introduced from Theodore Roosevelt National Memorial Park in 1964. This population was tested thereafter; calves were vaccinated, and all seropositive animals were removed. The last identified reactor was removed in 1967, and all adult bison tested negative in 1968. Late in 1968 and in 1969, some bison escaped from the wildlife park, and attempts were made to return them to the park. By 1970, however, nine bison were free-roaming because they could not be recaptured. The herd subsequently grew in numbers (Peterson et al. 1991b). About 1980, the animals began to winter on the NER, where they came into contact with winter-fed elk that were known to be infected with brucellosis. Cattle were not present on NER. In 1989, 11 of 16 bison collected on NER tested seropositive for brucellosis. On the basis of their modeling results, Peterson et al. (1991b) believed that the bison became infected in about 1980, and they noted that the bison herd first wintered on the NER, a potential source of *B. abortus* from winter-fed elk, in 1979-1980. Because the GTNP bison herd is isolated from the YNP bison herd by the continental divide, infection in GTNP bison is assumed to have derived from their contact with infected elk on the wintering grounds. Although the possibility of brucellosis having survived in the bison at the time of their escape from the wildlife park cannot be ruled out, transmission from elk seems more probable.

Two horses contracted brucellosis in the Jackson, Wyoming, area, where the only known source of the disease was elk on the winter feeding grounds (see p. 35, "True Prevalence").

One of the most contentious issues—because it is key to determining the need for control of the disease in GYA wildlife—is the probability of transmission of brucellosis between free-roaming bison and domestic livestock. Nearly all parties to the controversy agree that the risk of transmission of brucellosis from bison to cattle in the GYA is small, but not zero. Defining small depends on whether transmission has occurred in the past and, if so, how often. That is key to determining the need to control brucellosis in bison. Advocates of no control maintain adamantly that no case of transmission of brucellosis from bison to cattle in the free-roaming state in the GYA ever has been documented. Advocates of the need to control the disease in bison to protect livestock in the surrounding areas maintain equally stoutly that there is clear epidemiologic evidence that transmission from wildlife has occurred at least six times in the recent past, two of which might have been due to bison.

EPIDEMIOLOGIC EVIDENCE OF
TRANSMISSION FROM WILDLIFE TO CATTLE

The differing interpretations of epidemiologic evidence on the two sides of the issue are the crux of the controversy. This evidence is summarized in a field report submitted to APHIS in December 1996. Between about 1961 and 1989, cattle on six ranches in the GYA became seropositive for brucellosis after testing brucellosis-free. One of the ranches was east and five were west of the continental divide in the Jackson Hole region. The evidence consisted of seropositive tests for brucellosis in cattle herds in which the disease had not previously been present, and no known source of infection occurred in cattle in the local area or in stock imported to the properties. On each of five ranches, a single outbreak occurred. On the sixth ranch, brucellosis appeared in a cattle herd in about 1961 (the exact date is not known); it was thought to have been eliminated, and the herd was found again to be seropositive when retested in 1969. One outbreak in 1988 and another in February 1989 (Cariman 1994) led to a court case in which the Parker Land and Cattle Company sued the U.S. government for damages for failing to control elk movements from the NER to private lands (Parker vs. U.S.A. and Peck vs. U.S.A. 1992). The court concluded that the brucellosis outbreak was most likely caused by contact with infected elk or bison but the plaintiffs failed to prove that the elk or bison came from the NER, GTNP, or YNP. Several elk winter feeding grounds operated by the Wyoming Game and Fish Department are between the Parker ranch and the NER. No outbreak of brucellosis in cattle in that problem area has been reported since 1989. Cattle producers in the GYA routinely vaccinate their herds for brucellosis. Vaccination is required in Idaho and strongly recommended in Montana and Wyoming.

In four of the cases, anecdotal evidence was provided that elk were adjacent to or moving onto the property; the other two cases included anecdotal evidence of elk and bison presence. Most of the elk were associated with various winter feeding grounds on which elk concentrations foster transmission of *B. abortus*. Free-roaming elk herds, thought at the time of the first reports not to carry brucellosis, were found on further testing to have a relatively high proportion of seropositive individuals. By 1977, brucellosis had been detected on feeding grounds (Thorne et al. 1997). The bison in both cases would have come from the GTNP herd, in which 69% of individuals tested in 1988-1989 were seropositive for brucellosis (Peterson et al. 1991a,b; Williams et al. 1993, 1997).

Those six cases of purported transmission of brucellosis from wildlife to cattle are based on circumstantial evidence. The facts were derived from field operations of the federal-state cooperative program to eliminate brucellosis from domestic cattle in the United States. The data never were intended to meet the standard of scientific research, and inconsistent record retention resulted in further gaps in the documentation. The cases were summarized after the fact, some without supporting documents, which were discarded in the meantime. The only thing definite is that cattle in the herds tested seropositive for brucellosis. Assuming that elk and bison were in contact with cattle, there is no way to determine whether they were infective at the time and whether opportunity for transmission presented itself. Similarly, the possibility of infection from cattle is difficult to eliminate entirely, because it is always hard to prove that an event did not happen.

Some observers have noted that in states that have eliminated brucellosis from cattle in the past, occasional outbreaks are typical for some time after a state has been declared class-free by APHIS. That is because the disappearance function of the disease does not decline to zero at a constant rate but rather has a tail of gradually decreasing probability. Given the pattern of outbreaks in cattle in the GYA, with no new cases since 1989, this area might simply be mimicking the temporal pattern observed elsewhere where transmission from wildlife was not an issue. Or it could be maintained that the lack of outbreaks since 1989 is attributable to diligent cattle vaccination by ranchers. Given the ambiguity allowed by epidemiologic evidence in this situation, wildlife cannot be determined to be the source of brucellosis infection in these six cases.

BISON AND ELK BEHAVIOR AND TRANSMISSION

Considerable caution should be exercised in extrapolating results from cattle to bison beyond the consideration of a long, separate evolutionary history. There are fundamental differences between how cattle are managed and the natural behavior of free-roaming bison in the GYA. First, domestic bulls are placed with cows in lower relative numbers (typically 1:20 to 1:30) than the sex ratios of unmanipulated bison of about 1:1, or slightly skewed toward females (Meagher 1973; Van Camp and Calef 1987; Berger and Cunningham 1994). Second, domestic bulls are placed with cows only during the breeding period, then removed; the bison sexes can intermingle throughout the year. Third, courtship is perfunctory in domestic stock because the highly skewed sex ratio largely eliminates male-male competition. Bison males compete strongly for females, and dominant bulls form close "tending bonds" with estrous females that last several days, during which the male is never more than several meters away from the female. Younger males might maintain tending bonds with females at an earlier stage; thus, females can have multiple consort males in close attendance before and leading up to breeding. The chance of nonvenereal transmission between the sexes is increased because of this protracted courtship behavior.

Still, the two most probable sources of *B. abortus* transmission are abortion or birth when infectious materials are in the environment. Because of long exposure of bison to *B. abortus*, they respond to it more like chronically infected cattle herds in which selection for genetic resistance has occurred. In about 75 years, only four cases of abortion in YNP have been recorded (Rhyan et al. 1994); of course, regular surveillance is impossible given the large numbers and scattered distribution. The real number, therefore, has to be greater. But if abortion were common, many more cases would be expected to have been reported. In two cases, abortion sites remained culture positive for *B. abortus* for at least 2 wk (J. Rhyan, USDA, pers. commun., 1998).

Abortion among elk on the NER and Wyoming Game and Fish Department feeding grounds has been estimated at 7% (Smith and Robbins 1994) to 12.5% (Herriges et al. 1991) of pregnancies. Given such a high abortion rate and the high concentration of animals, transmission is highly likely. Indeed, Thorne et al. (1997) suggest that any elk that lives a long life and winters on a feeding ground is likely to become infected.

Also important is the difference in probability of association between elk and bison and cattle. Elk usually move away from areas used by cattle (Skovlin et al. 1968; McCullough 1969; Oakley 1975; Long et al. 1980; Mackie 1985), and this would reduce the contact between the two species. Bison, in contrast, are behaviorally dominant over cattle and respond to them aggressively if they approach within 5 m (Van Vuren 1982). However, they tolerate them when in proximity, and in one case, Van Vuren (1982) observed a domestic cow that joined a bison social group for 7 days.

In normal birth, the probability of transmission of *B. abortus* to cattle is influenced by the birthing behavior of bison and elk. Wild ungulates are categorized by birthing behavior as *hiders* or *followers* (Lent 1974). Hiding and following are major strategies used by mothers to avoid predation on their offspring. Hiders use dispersion, crypsis, and concealment to prevent discovery of offspring by predators, whereas followers depend on precocial offspring (offspring that can stand soon after birth), which can run with the mother to escape predators. Hiding is characteristic of species that have access to concealment cover in their habitat. Following is characteristic of herding species; herding is usually associated with open habitats that lack concealment cover and is itself a strategy for countering predators (McCullough 1969; Hamilton 1971).

Elk are classic hiders (Geist 1982). Females approaching labor isolate themselves from the herd (often moving several kilometers away) and seek cover in vegetation or broken terrain to give birth (Johnson 1951; McCullough 1969). After giving birth, the cow meticulously cleans the site (Livezey 1979; Clutton-Brock et al. 1982) and then moves the calf several hundred meters away to hide (Altmann 1952; Clutton-Brock et al. 1982).

The sanitation of the birth site by the mother is thorough. Females search the ground and consume small bits of birth tissue (Livezey 1979) and grass stained by fluids (Clutton-Brock et al. 1982). Bauer (1995) reported, "As we watched, the cow not only devoured the placenta and birth membranes, but also seemed to be eating the earth and grass that were saturated with birth fluids." Fraser (1968) noted that hiders eat afterbirth materials more for protection of the young than for physiologic reasons and that removal of vegetation and soil would remove any traces of scent from the site. Indeed, the entire suite of behavior of the elk cow and calf at birthing is linked to concealing the presence of the calf from predators. The calf hides alone while the cow feeds or beds in the vicinity, returning only long enough to nurse (McCullough 1969). The mother licks the calf's perineum during suckling; this stimulates voiding, after which she ingests the feces and urine (Arman 1974). The hidden calf remains motionless if approached during the first 3 or 4 days of life, running only at the last instant if hiding fails; the cow defends the calf from predators (Murie 1951; McCullough 1969). The cow and calf usually rejoin the herd in 2 or 3 wk after birth (Altmann 1952, McCullough 1969).

The evolution of antipredator behavior in elk has resulted fortuitously in behavior that reduces the likelihood of *B. abortus* transmission. The dispersed birthing area and sanitation of the birth site result in a low probability that other animals will come into contact with infectious birth products.

The consensus of respondents to the National Research Council questionnaire was that *B. abortus* is not self-sustaining in elk herds that are not concentrated on winter feeding grounds. That is cited as the reason that the elk in the northern Yellowstone herds that are not winter-fed have a seropositive rate of only 1-2% (M. Meagher, USGS., pers. commun. as cited by Smith and Robbins 1994; Rhyan et al. 1994), whereas those using winter feeding grounds in the southern part of the GYA have an average seropositive rate of 34%. A somewhat higher seropositive rate (5 of 126, or 4%) in northern-range elk was reported by Thorne et al. (1991), but this could reflect sampling error.

In contrast with elk, bison offspring are followers, as is consistent with the highly developed herding social structure in this species (McHugh 1958; Meagher 1973; Lott 1974). Pregnant females separate from nonpregnant females to form nursery herds (McHugh 1958; Lott and Galland 1985; Meagher 1986; Berger and Cunningham 1994). Females give birth either alone or in small subgroups and might seek cover, depending on what is available in the environment occupied by the nursery herd at the time of birth (McHugh 1958; Lott and Galland 1985). Nevertheless, birth occurs either in or close to the herd. Mean time from birth to standing by the calf is about 11 minutes and from birth to nursing about 32 minutes (Lott and Galland 1985). The mother usually consumes the afterbirth (McHugh 1958; Fraser 1968; Lott and Galland 1985; J. Berger, U. Nev., pers. commun., 1997). However, detailed observations of how thoroughly the site is cleaned are not available. In bison, consuming the afterbirth might be related mainly to hormonal and physiologic needs; the antipredator benefits of consumption would seem minimal in a species that lives in large herds in open areas and has offspring that are conspicuously different from the adults. For bison calves, the major antipredator protection is the herd. Calves are protected from predators not only by their ability to run with the herd, but also through defense by the large, formidable mothers, whose common inter-est—protection of young from predators—presumably is the selective advantage of forming separate nursery groups in the first place. If consump-tion of the afterbirth in bison is related to hormonal factors rather than predator avoidance, it might be that the birth site is not so well sanitized as by elk.

Giving birth within the herd concentrates the afterbirth in space and increases the likelihood of encounters with other herd members and roving males. That increases the probability of transmission of B. abortus associated with birth products among bison and to other species that might accidentally or purposefully encounter the nursery herd area. The dispersed distribution of birthing in elk, in conjunction with their thorough cleansing of the site, makes the probability of transmission of B. abortus among elk or from elk to other species, lower than for bison.

Abortion by B. abortus-infected females is a more serious risk factor for disease transmission than is normal birth. Abortion is spontaneous and typically occurs in the third trimester of pregnancy. That timing places most abortions in the winter when both bison and elk are concentrated, some on artificial feeding grounds. Abortion occurs out of synchrony with the social structures of normal birth and decoupled from the usual entraining of endocrine activity that regulates normal birthing behavior. Concentration of animals on winter feeding grounds or, in bison, by the natural herd structure greatly increases the potential for contact with aborted fetuses and other afterbirth products. In addition, disruption of normal hormonal controls results in retention of placentae in bison and failure of the females to clean up the birth products. Retained placentae in bison can attract the attention of other herd members and roving bulls and extends the exposure period of *B. abortus* in time and space. Elk apparently do not retain the placenta after abortion, and they can reach it and remove it before it hits the ground (Thorne et al. 1978, 1997). In their study of penned elk, Thorne et al. (1978) reported that aborting females attempted to eat their fetuses but that they might have been only partially consumed. In this captive herd, other females were observed to investigate and lick the partially expelled fetuses during abortion. Intact fetuses and afterbirth remaining at the abortion site would greatly increase the probability of transmission between animals. Furthermore, at the typical time of abortion, winter temperatures and moisture would favor survival of *B. abortus* in the environment, as would sequestration of *B. abortus* in larger masses of birth tissue not consumed by the female.

TRANSMISSION BY OTHER SPECIES OF UNGULATES

Other wildlife species have the potential to contract and transmit brucellosis (see review of Remontsova 1987). Other wild ungulates in the GYA—mule deer, white-tailed deer, antelope, and bighorn sheep—have never been documented to harbor the microorganisms (McKean 1949; Steen et al. 1955; Shotts et al. 1958; Trainer and Hanson 1960; Rinehart and Fay 1981; Jones et al. 1983; Gates et al. 1991; K. Aune, Mont. Dept. Fish, Wildlife, and Parks, pers. commun., 1997). Moose are known to contract the disease, although moose living in an area where cattle were heavily infected by *B. abortus* tested seronegative (Hudson et al. 1980). None of several dozen moose tested in the GYA was seropositive (T. Thorne, Wyo. Game and Fish, pers. commun., 1997). Moose are considered a dead-end host for brucellosis and are not thought to be a threat to transmit the disease. They do not seem to be involved in the epidemiology of brucellosis. Moose are typically solitary, and yet the rare occurrence of brucellosis in moose, a species that does not usually carry or perpetuate the disease, illustrates the possibility of transmission of *B. abortus* among the species that do. Surveillance for the disease in moose or other wildlife species that are dead-end hosts might be a way of estimating the probability of rare events of transmission among bison, elk, and cattle.

POTENTIAL ROLE OF CARNIVORES IN TRANSMISSION

Predators can become infected with *B. abortus*, and they are potential reservoirs for transfer to other species. The most thorough work on *B. abortus* in carnivores is the study done on coyotes by Davis et al. (1988). They fed macerated cattle fetal material infected with *B. abortus* to 40 brucella-negative coyotes, and 32 became seropositive. They also found that *B. abortus* can pass through the digestive tract of coyotes and remain viable in feces and urine. In each of four trials, 10 exposed coyotes were put in 1-hectare pens with six uninfected heifers. *B. abortus* transmission occurred in three heifers in one trial, and they aborted. No transmission occurred in the other trials; 3 of 24 heifers were infected overall. The heifers probably became infected through contact with urine or feces of coyotes (D. Davis, Texas A&M, pers. commun., 1997). Coyotes can potentially serve as a bioassay for *B. abortus*; a survey of two-thirds of the counties in Texas showed that seropositivity in coyotes corresponded to the known distribution of brucellosis in cattle (D. Davis, Texas A&M, pers. commun., 1997).

Transmission in the Davis et al. (1988) study occurred under confinement at artificial densities of both coyotes and cattle. Although it does verify the possibility of transmission, that cannot be translated into probabilities of transmission under natural range conditions.

Carnivores of YNP—including grizzly bears, black bears, wolves (*Canis lupus*), coyotes, and foxes (*Vulpes fulva*)—are known to contract brucellosis (Zarnke 1983; Remenëtïsova 1987; Morton 1989; Johnson 1992), presumably through consumption of infective tissues during predation and scavenging. Of 122 grizzly bears tested in Alaska, six were seropositive (Zarnke 1983). Current estimates of grizzly bear population size in the GYA are around 300 (Eberhardt and Knight 1996). There were an estimated 650 black bears in the GYA in the late 1970s (Cole 1976), but their numbers might have declined (Schullery 1992). YNP has no current estimate of black bear numbers; they are considered common in the park (Gunther 1994), but they are seldom mentioned with reference to brucella transmission. Wolves were extirpated from the GYA by the early 1930s and have been reintroduced only recently (Weaver 1978; Yellowstone Science 1995; Bangs and Fritts 1996). Consequently, the ecosystem role of wolves has been missing for many years and is only now being re-established. The current wolf population is about 100. Coyotes are ubiquitous in the GYA.

Any debilitation due to brucellosis (Tessaro 1987; Thorne et al. 1997) would predispose adult elk and bison to predation. Grizzly bears, wolves, and coyotes scavenge and all are predators on calves. Scavenging makes them vulnerable to contact with products of birth and abortion, the likely route of acquisition of *B. abortus*, but it is highly unlikely that these species directly transmit the bacterium back to ungulates. They are considered dead-end hosts. Transmission of *B. abortus* by carnivores through transport of infective materials from birth or abortion sites to other areas, however, is a concern.

Carnivores could have positive and negative effects on the dynamics of *B. abortus*. On one hand, by consuming products of birth and abortion they remove the bulk of infectious materials from the site and expose remaining *B. abortus* on the soil and vegetation to light and desiccation, to which they are vulnerable (Mitscherlich and Marth 1984). Although it has not been quantitatively documented, known carnivore behavior makes the existence of a healthy complement of predators almost certain to be a major factor in reducing the probability of *B. abortus* transmission within the wildlife community and between wildlife and domestic stock. Predation and scavenging by carnivores likely biologically decontaminates the environment of infectious *B. abortus* with an efficiency unachievable in any other way.

On the other hand, carnivores might contribute to transmission probabilities by transporting infectious materials from one site to another. Particularly troublesome is the possibility of transporting such material between exclusive wildlife and cattle areas kept geographically separated by management. No data are available to address this question directly; the potential risk must be evaluated on the basis of what is known about the behavior of these carnivores.

Ordinarily, urine and feces from predators would be unlikely routes of direct transmission of *B. abortus* because the number of organisms shed is small in relation to the infective dose for cattle (Morton 1989), and cattle, bison, and elk would not be attracted to or likely to come into contact with them accidentally. However, one exceptional circumstance should be noted. *B. abortus* apparently can pass through the gastrointestinal tract of predators and survive in their feces (Davis et al. 1988). Under some conditions of mineral deficiency, domestic cattle show depraved appetite, or pica, in which they consume a variety of atypical objects (Church et al. 1971). Similarly, wild ruminants commonly visit mineral licks and consume soil during some times of year, usually during periods of rapid growth in the spring. Rodents and rabbits are well known to consume bones and antlers, presumably for the minerals in them. In a mineral-deficient area in south Texas, cattle were observed to consume coyote droppings (D. Davis, Texas A&M, pers. commun., 1997), which commonly contain small mammal bones, a mineral source. Also, reindeer penned with foxes consumed fox feces (Morton 1989). If such behavior occurred in a brucellosis area, the probability of transmission of *B. abortus* from predators to herbivores could be substantially increased. Whether such behavior occurs in bison, elk, or cattle in the GYA is unknown.

A more important concern with predators is their transport of infected ungulate-carcass materials from a death or abortion site to other areas. Internal organs of large animals are usually consumed first, and skeletal muscle and other body parts later (E. Gese, NWRC, Ft. Collins, Colo., pers. commun., 1997). Heads, bones, and other hard materials are consumed last or not eaten at all. Coyotes and wolves sometimes transport pieces of carcasses short distances to nearby preferred microsites to complete consumption, but this would spread the bacteria only locally and not greatly increase the likelihood of transmission. Grizzly and black bears are not known for transport of carcasses or parts from the site of death; they do not usually move carcasses elsewhere to cache them, although they sometimes cover the carcass at or near a kill site (Craighead et al. 1995), which might preserve *B. abortus* for longer periods. They usually feed on site. Bears are followed by dependent offspring and do not provision.

Longer-distance transport could occur as a result of caching carcass parts and provisioning pups sequestered in dens; these behaviors are shared by coyotes, wolves, and red foxes. Parts of carcasses carried by mouth (usually pieces containing bones, which afford structural integrity) can be transported great distances. Soft tissues may be consumed and subsequently regurgitated at the den. Caching has been reported in wolves (Murie 1944; Mech 1970; Harrington 1981), coyotes (Weaver 1977), and especially red foxes (Vander Wall 1990). Caching—thought to be a way to extend the time that food is preserved, to protect it from competitors, and to hedge against difficult hunting times—seems to be most common in populations with smaller home ranges and greater population densities. In Alaska, wolves often disperse and cache chunks of caribou, burying them in soil or in creeks covered with moss (K. Taylor, Alaska Dept. of Fish and Game, pers. commun., 1997). Recently in YNP, wolves were observed to kill a pronghorn fawn and cache the carcass (F. Camenzind, Jackson Hole Conservation Alliance, pers. commun. 1997). Foxes, which have relatively small home ranges, cache frequently, coyotes less commonly, and wolves least commonly. The potential for *B. abortus* transmission by red foxes (Johnson 1992) should be considered more carefully, given their well-developed caching behavior (Vander Wall

1990), their common occurrence in YNP (Gese et al. 1996a), and their caching frequency (E. Gese, NWRC, Ft. Collins, Colo., pers. commun., 1997). Their small home ranges would limit long-distance transport, but their active caching would increase the local foci of B. abortus contamination.

Usually pieces cached are relatively small, weighing a few kilograms or less. Ordinarily, elk and bison calves are too large to be cached whole. But aborted fetuses are smaller. Abortion can occur at any time during gestation, but it is most frequent in the late second and early third trimesters (D. Davis, Texas A&M, pers. commun., 1997). Weights of 37 elk fetuses aborted in captive studies averaged 7.9 kg (range, 4.5 to 11.4 kg) (W. Cook, U. Wyom., pers. commun., 1997). Bison fetuses would ordinarily be even heavier. Consequently, predators are likely to reduce fetuses to smaller parts before caching, as they do with normal-birth calf carcasses.

A considerable amount of information on caching behavior in coyotes has been obtained inadvertently through radiotelemetry studies of survival of young ungulates. From radiotelemetry studies of antelope fawns, coyotes have been found to cache carcass parts commonly in Arizona (R. Ockenfels, Ariz. Game and Fish Dept., pers. commun., 1997), Kansas (E. Finck, Emporia State Univ., pers. commun., 1997), and Oregon (R. Cole, Hart Mtn. National Antelope Refuge, Ore., pers. commun., 1997). Coyotes also have been found caching black-footed ferrets in Wyoming (E. Williams, U. Wyom., pers. commun., 1997). In the GYA, Weaver (1977) and E. Gese (NWRC, Ft. Collins, Colo., pers. commun., 1997) have observed caching in the snow by coyotes. In GTNP, Weaver (1977) observed coyotes caching meat from an elk carcass. He observed 16 snow caches, ranging from 10 m to more than 400 m from the carcass, one of which contained a 78-g piece of meat. A coyote often would bed down on a cache site, perhaps to mask the site from competitors. In YNP, Gese (NWRC, Ft. Collins, Colo., pers. commun., 1997) observed coyotes caching elk and bison (less commonly because bison mortalities are far fewer than those of elk), usually less than 300 m from the carcass. Most sites were cleaned out fairly soon thereafter. Despite their potential to contract brucellosis, Gese et al. (1997) reported that 70 blood samples of coyotes from Lamar Valley in YNP were all seronegative. Caching in Lamar Valley was done most frequently by red foxes, however.

To prevent discovery by competitors, carnivores completely bury caches several inches below the soil and conceal the sites. Large herbivores are extremely unlikely to encounter these sites or, if they do, to come into contact with the buried materials. In addition, the parts that are cached are usually muscle and bone, which are less likely to harbor B. abortus than reproductive tissues are. Chance of transmission of B. abortus to cattle, bison, or elk would seem unlikely with earthen caches. Caching in the snow during the winter, however, as has been reported for coyotes in GTNP (Weaver 1977), presents greater hazards. Furthermore, foxes steal small pieces, including soft tissues, from carcasses being fed on by coyotes and bury them in the snow as well (S. Grothe, Mont. State Univ., pers. commun., 1997). Burial in snow would favor longer survival of *B. abortus*; if the cache were not retrieved, buried remains would be exposed on the surface of the ground when the snow melted.

Provisioning of pups, through transporting muscle and bone by mouth and regurgitating soft parts, is a more common predator behavior than caching and therefore more likely to move *B. abortus* from one place to another (and between geographically separated wildlife and cattle areas). E. Gese (Fort Collins, Colo., pers. commun., 1997) observed coyotes provisioning up to several miles from a carcass. The work of Davis et al. (1988) showing that *B. abortus* can survive in coyote urine and feces indicates that it is likely to remain viable in the partially digested stomach contents regurgitated for pups. These contents can include organ tissues that are most likely to contain *B. abortus*. Both members of the territorial mated pair and associated pack members ("helpers") engage in provisioning of pups (Hatier 1995). According to research on survival of *B. abortus* in the environment, the bacteria likely would survive during the transport of such materials and potentially could contaminate new sites that are far removed from the initial source. However, regurgitated material is consumed by the pups immediately.

To protect pups, den sites are usually hidden in rocky, rough, timbered areas where large herbivores are less likely to go. In addition, prey species probably avoid approaching predator dens. Wolves and coyotes switch den sites regularly; if *B. abortus* persisted long enough, herbivores could come into contact with it around abandoned den sites in the absence of predators.

The seasonality of provisioning pups by coyotes and wolves is important in relation to the time that *B. abortus*—principally through abortion and calving—might be in the environment. Timing of reproductive events in bison, elk, wolves, and coyotes in YNP is shown in Figure II-1. The major period of provisioning by canids in YNP occurs after the main period of birthing in bison but overlaps with elk calving. However, there are late, aseasonal births in bison (Meagher 1973; Berger and Cunningham 1994; Roffe, unpubl. data) that are possible sources of *B. abortus* during the provisioning periods of wolves and coyotes. Such aseasonal births are also known in elk (Smith 1994). Similarly, *B. abortus* could be acquired by consumption of infected tissues from whole carcasses during predation at any season.

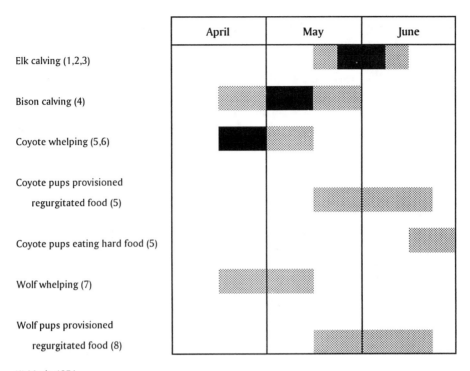

(1) Murie 1951
(2) Houston 1982
(3) Taber et al. 1982
(4) Meagher 1973
(5) Hatier 1995
(6) Eric Gese, pers. commun.
(7) Yellowstone Science 1995
(8) Extrapolated from whelping period

FIGURE II-1. Timing of reproduction events in bison, elk, coyotes, and wolves in YNP showing degree of overlap of *Brucella* exposure during normal birth in bison and elk, and provisioning periods in wolves and coyotes.

The possibilities of transfer of *B. abortus* between wildlife species become smaller with each ecologic complication. The wolf, coyote, or fox needs to be in contact with *B. abortus* via tissues from bison or elk; the *B. abortus* must survive being transported to a cache or den site; and it must persist at the new site long enough and in sufficient numbers to infect a susceptible bison, elk, or domestic cow that encounters the den site more or less through random movement. The chance of transfer of *B. abortus* among elk, bison, and cattle through the activities of predators and scavengers seems extremely small under most conditions.

Transmission of B. abortus among ungulates indirectly through predators and scavengers cannot be completely ruled out, but it seems unlikely in comparison with transmission by direct contact among bison, elk, and cattle. On balance, the positive advantage of sanitary activities of large carnivores in the GYA greatly outweighs the negative effects of their possible role in transmission of B. abortus. Consequently, the presence of carnivores in the ecosystem probably reduces the frequency of B. abortus moving between bison, elk, and cattle.

ROLE OF OTHER WILDLIFE SPECIES

A wide array of other wild vertebrates, insects, and other arthropods that occur in the GYA can harbor B. abortus (Remenëtìsova 1987), including rabbits (Peterson 1991), rodents (Thorpe et al. 1965; Moore and Schnurrenberger 1981; Rinehart and Fay 1981), and ticks. Nevertheless, these species are incidental hosts and are not important in the transmission of brucellosis.

BISON MOVEMENT OUT OF YELLOWSTONE NATIONAL PARK

There is no risk of transmission of B. abortus from bison to cattle in the northern range if bison do not leave YNP. Cattle grazing is not permitted inside YNP. For many years, bison have been reported to move out of YNP in hard winters (Meagher 1973, 1989), and a large kill of bison (1,084 animals) outside the park during the winter of 1996-1997 produced a great controversy (Peacock 1997) and reinvigorated the debate about brucellosis.

Weather and Bison

The effect of winter weather on bison movements outside YNP boundaries is a topic that is amenable to modeling, but little effort has been given to it. The first issue is the frequency of hard winters in the GYA. Farnes (1996) presented 45 years of climatic data (from the winter of 1948-1949 to the winter of 1992-1993) from several stations around the northern range, and he has furnished comparable data from 1993-1994 to 1996-1997 (P. Farnes, Snowcap Hydrology, Bozeman, Mont., pers. commun., 1997) for a total time series of 49 years. He presented the measured values and calculated severity indexes based on a range from -4 (severe) to +4 (mild). Variables included were snow water equivalent in inches (SNOW) from Lupine Creek and Crevice

Mountain snow courses (2,249 m and 2,560 m, respectively), cumulative sum of minimal daily temperatures below -18°C (TEMPERATURE) measured at Mammoth (1,890 m), and summer rainfall (RAIN) measured as the sum of June and July rainfall in the previous year at Mammoth. He also presented a combined winter severity index that included a weighted measure of SNOW (40%), TEMPERATURE (40%), and RAIN (20%). Farnes (1996) noted that early-summer rains might be related to fall animal condition and likelihood of surviving the winter, as previously reported by Meagher (1973). The objective of our analysis, however, was to examine the influence of winter weather on bison movements independently of animal condition, so SNOW and TEMPERATURE were emphasized. RAIN was included in the analysis, but was not significant, either alone or through its contributing to winter severity index, so it is not discussed further.

Examination of the frequency distribution of the two measured winter variables showed that TEMPERATURE was skewed (mean, 178.8; C.V., 0.59; median, 154.0; skewness, 0.83), with most winters being on the mild side and less-frequent severe winters, whereas SNOW was normally distributed (mean, 21.1; C.V., 0.265; median, 21.3; skewness, -0.05); that is, most years are near the mean, but for extreme years, mild and severe winters are equally likely. Autocorrelation and cross-correlation with year showed no significant pattern over time for TEMPERATURE or SNOW, although there was a weak trend toward an oscillatory pattern for Farnes's (1996) snow index. Overall, the results suggest that hard winters occur roughly randomly, and a much longer time series would be required to detect any periodicity.

The second issue is whether the measured variables or indexes correlate with the numbers of bison moving out of YNP. Bison-population estimates over time are shown in Figure II-2. (Data on bison populations are from Yellowstone National Park (1997), and data on bison removals are from Dobson and Meagher 1996 and Yellowstone Science 1997.) Three periods of different management policy give different results. An early period from 1902 to about 1930 was a time of population recovery from low numbers that survived uncontrolled market hunting. Beginning in the 1930s and extending until 1967, large removals inside YNP controlled bison numbers. The natural-regulation policy was implemented after 1967, and an increase in numbers occurred, with removals consisting of animals moving outside YNP boundaries. Virtually all the bison moving out of YNP were killed and so were lost to the population. The issue is addressed through regression analysis on data after the implementation of natural-regulation policies in 1967. The effects of individual weather variables and indices and estimated bison population size on bison moving out of YNP are examined by simple and stepwise multiple regression.

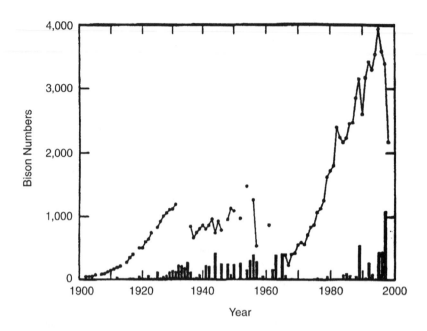

FIGURE II-2. Plot of estimated bison population (circles) and bison removals (bars) by year for YNP. Data on bison population are from Yellowstone National Park (1997: Appendix B) and bison removals from Dobson and Meagher 1996, Yellowstone Science 1997.

None of the weather variables or indexes shows a significant correlation with bison moving out of YNP; indeed, none is even suggestive. Only estimated bison population size is significantly related to the number of bison migrating out of the park (P < 0.001). The plot of bison moving out of YNP on estimated bison population, however, shows the relationship to be highly nonlinear (Figure II-3); indeed, the abrupt transition is best described as a threshold. Above a population of 3,000, bison show the greatest probability of moving out of YNP. Log transformation of bison moving out of YNP yields a significant linear fit with bison population ($R^2 = 0.53$; P < 0.001), but even that transformation does not adequately reflect the abruptness of the threshold.

To examine the effects of weather on populations below the threshold, the regression analysis is repeated for bison population estimates of less than 3,000. Again, no weather variable or weather index is close to significant. Cross-correlation shows that relationships are not delayed; zero lag yields the highest correlations. Again, bison population size is the most important variable, but it did not quite reach statistical significance (P = 0.06).

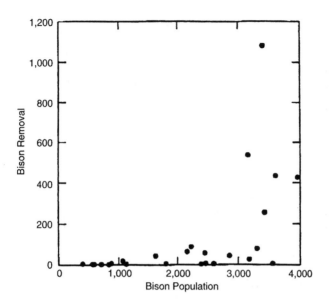

FIGURE II-3. Plot of bison removals on estimated bison population size for years under management by natural regulation (1968-1997).

The final issue addresses relationships at populations above the 3,000 threshold. One might expect the response to winter conditions to be strongest in large populations. Given bison populations of more than 3,000, does winter severity influence the number of bison moving out of the park? Regression analysis of bison populations on various indexes of winter severity in years when there were more than 3,000 bison show that SNOW and snow index are strongly related to bison moving out of YNP (Figure II-4) ($R^2 = 0.84$ and $P = 0.001$, and $R^2 = 0.71$ and $P = 0.009$, respectively). No other winter-severity variable is close to significance, nor does stepwise regression result in an improved fit.

Of the two measures of winter severity available, TEMPERATURE and SNOW, only SNOW proves to be important statistically in explaining bison movements. Figure II-4 shows that for populations over 3,000, the number of bison moving out of YNP increases rapidly with increasing SNOW (on average, 68 bison for each inch of SNOW). Furthermore, on the average, no bison moved outside YNP when SNOW was 17 in. or less. That average fails to capture the fact that historically some bison have moved outside the park even when the population was low (Meagher 1973). Nevertheless, 17 in. of SNOW is a useful benchmark for increased probability that bison will move

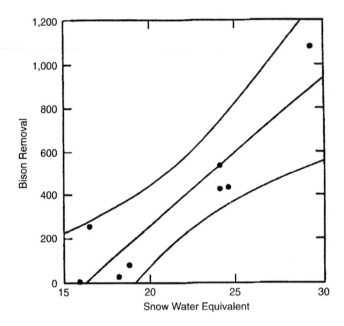

FIGURE II-4. Plot of bison removals on snow water equivalent for bison populations greater than 3,000. Lines are least-squares fit (Y = -111.16 + 68.40(X) and 95% confidence intervals.

out of YNP and, if not controlled, potentially come into contact with cattle. The two points in Figure II-4 farthest above the line (1991-1992 and 1996-1997) are identified by the statistical program (SYSTAT: SPSS, Inc. 1996) as outliers; this suggests the involvement of additional variable or variables. Many observers have noted that snow depth itself might be unimportant because bison are adept at digging craters in deep snow to forage. A freezing and thawing pattern that produces ice layers in the snow might be more important (Peacock 1997). In the winter of 1991-1992, an early snowfall was followed by a thaw, which in turn created an ice layer at ground level (Gese et al. 1996b); in the winter of 1996-1997 ice layers in the snow pack formed a physical barrier to foraging beneath the snow (Peacock 1997). Thus, relatively mild winters that have thawing followed by freezing might be more difficult for bison (and probably elk, which also dig craters in snow to forage) than severely cold winters with deep snow.

Farnes (Snowcap Hydrology, Bozeman, Mont., pers. commun., 1997) suggests that bison typically move to areas that have less than 6 in. of snow water equivalent and have available forage. He notes further that bison-snow relationships are complex, involving snow water equivalent, snow density, spatial and temporal variation, and perhaps other variables. Learned behavior and forage availability also interact with snow conditions to influence bison movements.

SNOW shows a normal distribution, and most values are expected to fall around the mean of 21.1 in. The winter carrying capacity of YNP is about 3,000 bison; this analysis suggests that above this population size, bison will move out of the park in all but the mildest winters (Figure II-4). Therefore, the regression in Figure II-4 would predict that under average conditions, about 332 bison will move out of YNP each winter, more in high-SNOW years and few or none in low-SNOW years. Experience with habitats outside of YNP might encourage bison movement in the future that is not driven solely by population size and winter severity. However, so long as bison moving out of the park are removed from the ecosystem, this behavior will be discouraged. Furthermore, the relative role of experience is not clear, for the basic tendency of bison to move in the face of adverse conditions seems to be a primary motivation, and the landscape funnels them to lower elevations outside the park. Therefore, experience might contribute to the tendency to move, but it is probably not necessary to account for the behavior.

Obviously, more years of data will be needed to refine the interaction of bison numbers, SNOW, and bison movement out of YNP. The relationship of SNOW to bison leaving the park (Figure II-4) is based on few points and has wide confidence limits. Only a few measures of weather variables are analyzed, and those only from a few sites. They might not have captured the important aspects of a hard winter from a bison's point of view. A measure of ice layering, for example, would be valuable. Current analysis of spatial data on weather variables now being done by Farnes (Snowcap Hydrology, Bozeman, Mont., pers. commun., 1997) will better elucidate the causal relationships between snow characteristics and bison movements and expand on the correlation reported here. For the present, however, the importance of this analysis is the degree to which it reveals broad patterns. Bison population size appears to be the overwhelmingly significant variable controlling movement of bison out of YNP. Still, as long as bison are artificially controlled at or near the YNP boundary, the predictions of the number of bison moving out of the park in relation to SNOW should be sufficient for management because they are not allowed to move further. If bison are not artificially controlled, however, continued, more directed research using refined measurements of specific attributes of weather and its variance over time, space, and topography would go beyond simply predicting how many bison move out of YNP to how far they go and for how long. Refinement in recording and modeling weather measurements and more data on bison movements will refine the quality of predictions but are unlikely to alter these conclusions.

Natural Regulation in YNP Bison

As just noted, the expansion of the bison population in YNP appears to be the fundamental force pushing bison out of YNP, contributing both to increased risk of transmission of *B. abortus* to livestock and to the need to take action to deal with bison in unwanted places. Whether natural regulation is occurring is the underlying issue for *B. abortus* risk-management and bison-management policy: Is there evidence that bison are likely to reach a dynamic equilibrium with the carrying capacity in YNP?

That question can be addressed by examining the natural logarithm of bison population estimates over time (Figure II-5). If the population growth rate is constant, then logarithmic plots on time should be linear. In Figure II-5, it can be seen that the plots are approximately linear over much of the population growth record. The reduction in population growth rate toward the end of the early period (1902-1930) is accounted for largely by bison removal inside YNP (Figure II-2) and therefore does not reflect carrying capacity. Offsetting time by 51 years (open circles in Figure II-5) demonstrates that the growth trajectory of the early period is consistent with the growth trajectory of the natural regulation period (after 1967).

The reduction of population growth rate after about 1980 (Figure II-5) is more complicated. The population was reaching numbers at which SNOW becomes more influential in movements of bison out of YNP. Removal of these animals (Figure II-2) therefore accounts for at least part of the diminution in population growth. The last three winters, particularly, have been marked by the removal of many bison (Figure II-2). However, hard winters contribute to natural mortality inside YNP as well, and this is part of natural regulation. Estimates of natural mortality were not available over most of this time, so the effects of artificial removal, compared with natural mortality, are difficult to discern. For the last 3 years, the difference in population estimates between years is only partially accounted for by removals; this suggests a substantial residuum of natural mortality. Certainly, the combination of both resulted in a large reduction in bison numbers (3,400 to 2,169) over the winter of 1996-1997.

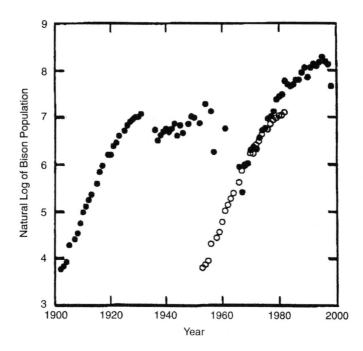

FIGURE II-5. Plot of $\log_e$ of estimated bison population on year for YNP bison (filled circles). The three time periods noted in Figure 2 can be seen here. The open circles are the data points for the early period (1902 to 1930) advanced by 51 years to align the early growth period with that following implementation of the natural regulation policy in 1967.

Nevertheless, the period 1972-1994 was one of relatively few removals (Figure II-2). Furthermore, when the effects of removals are canceled by adding them back to the following year's population estimate, a remarkably good linear fit of bison numbers on time is obtained for the estimates from 1972 to 1995 ($R^2 = 0.987$; $P < 0.001$) (Figure II-6, filled circles). The linear fit in Figure II-6 shows that the annual increment in the bison population was more or less constant at 145 per year. The close fit suggests that natural mortality in YNP during this period was low and roughly constant. That in turn suggests that natural mortality was minor in years other than those with hard winters and that at high populations a large portion of mortality was due to artificial removals when bison moved out of YNP.

Only the last 2 years of the time series (open circles in Figure II-6) deviate from the regression line, and even the point for 1996 is within the variance previously observed in the time series, which might be related to actual

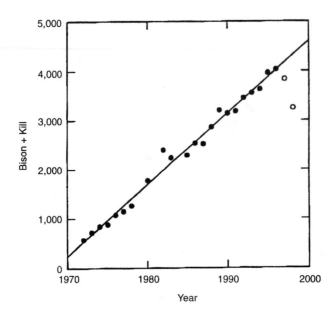

FIGURE II-6. Plot of estimated bison population plus previous year's removal on year for 1972-1997. The line represents a least-squares linear regression equation to the filled circles (1972-1995); $Y = -286137.746 + 145.371(X)$; $R^2 = 0.987$, $P < 0.001$. The open circles (1996 and 1997) show years in which annual increment was below the regression line. See text for further explanation.

changes in bison increment or to error in bison population estimation. Including all but the last year (1997) changes the relationship only slightly ($R^2 = 0.98$; $P < 0.001$), with the estimate of annual increment lowered from 145 to 143 bison. Only the point for 1997 clearly deviates from the linear regression (although inclusion gave an $R^2 = 0.955$ and $P < 0.001$) but lowers the annual increment to 134. The winter of 1996-1997 was a severe one that caused substantial mortality, both outside and in YNP. Even the previous two winters had above-average SNOW (24.6 and 24.1 in., compared with the average of 21.1 in.). Thus, even in the most recent years with high bison numbers, there is little evidence of natural diminution of bison population growth except that induced by the severe winter in 1996-1997.

Although the absolute annual increase is essentially constant at 145 bison, the per capita rate is declining. For example, the per capita growth rate of the population of about 530 in 1972 (at the beginning of this period) would have been 0.27, whereas that of the population of about 3,730 in 1994 (at the end of the period), would have been 0.04. This apparent contradiction can be explained by a simple economic analogy. Assume a constant, fixed income was always put into an account. As the account grows, the income becomes a lower proportion of the capital similar to the decline in per capita growth rate in a density-dependent population response. However, the income remains constant, so the capital continues to grow despite the relatively lower proportion of the capital the income represents. So long as the income remains fixed, the capital will continue to increase, and the per capita rate of growth will never drop to zero.

That result is unexpected because with increased population size, density dependence in ungulates is usually expressed by declines to zero in both rate of increase and absolute increase. There is no established model for this population behavior in ungulates, although it was anticipated by McCullough (1990). This kind of population behavior is known for other taxa. In territorial species, such as most passerine birds, and carnivores, such as wolves and coyotes, territorial holders are the only individuals that can effectively breed and only so many territories can be fit into the habitat. Thus, reproductive output tends to be relatively stable over broad ranges of total population size. We do not know what mechanism might be operating to stabilize the annual increment in YNP bison.

That the rate of increase declines, but not to zero because of a virtually constant absolute annual increment, suggests that some variable other than scramble competition (each individual doing the best it can to obtain resources) is modifying the density-dependent process. Because adult bison mortality seemed to be relatively low during the years included in this analysis (1972-1994), calf recruitment (birth and survival to yearling age) is the likely source of the constant absolute annual increment.

Estimates of pregnancy rates have shown that a relatively high proportion of adult cows become pregnant. From 1935 to 1950, up to 90% of adult females were pregnant (Coburn 1948; Rogers 1950). In 1988-1989, 74 of 102 (73%) of mature females were pregnant (Pac and Frey 1991); in 1990-1991, 54 of 68 (79%) were pregnant (Meyer and Meagher 1995). Aune (Mont. Fish, Wildlife, and Parks, pers. commun., 1997) reported that 90% of radiocollared bison cows calved in 1995-1996. Kirkpatrick et al. (1996), however, reported a 3-year (1990-1992) mean calving rate for the YNP northern-range herd of 52.6%. T. Roffe (USGS, pers. commun., 1997) found that 90% of a sample of 52 bison cows killed in the winter of 1996-1997 were pregnant. High pregnancy rates point to survivorship between midgestation and 1 year of age, rather than failure to conceive, as the critical variable in calf recruitment. Meyer and Meagher (1995) reported that about 400-600 calves were born each year before 1995. An annual population increment of only 145 would require that a large proportion of calves died, because adult mortality was low. The most likely explanation for the constant increment is that either dominance in females is determining success in recruiting calves so that some cows consistently produce while others consistently fail or there are few good habitats in which females succeed in recruiting calves, whereas elsewhere females fail. These two possibilities are not mutually exclusive—dominant females are likely to displace subordinates from the best habitat—so both might be interacting to result in constant annual recruitment.

YNP bison population behavior contrasts with the northern-range elk population behavior, in which the dynamic equilibrium is expressed by year-to-year changes. The lack of stabilization of bison population growth over time since the natural-regulation policy was adopted suggests that bison have expanded like a wave front across suitable habitat in YNP with little diminution until now they are pressing against the borders of YNP in winter. The prospect, therefore, is for the bison population to increase over some years until the coincidence of a high population and a hard winter results in the population being reduced once again (as happened in 1996-1997). Given the lack of a dynamic equilibrium, the bison numbers are expected to start building again. It will be instructive to determine whether the constant increment persists and the absolute value remains the same during recovery as during the 1972-1994 period.

McCullough (1990, 1992) proposed that ungulates that feed on homogeneous, relatively coarse, low-quality food (bulk feeds), as bison do, with diets 95% or more grasses and sedges (McCullough 1980; Van Vuren 1982; Singer and Norland 1996), might show roughly constant population growth until carrying capacity is reached, at which point growth drops abruptly to zero (a plateau with a cliff edge). That contrasts with animals such as elk (Kufeld 1973; Hobbs et al. 1979; Marcum 1979; McCullough 1980), in which qualitative aspects of the forage result in a declining (ramp) adjustment of population growth before carrying capacity is reached at zero growth. Environments that present considerable amounts of unoccupied habitat to which the population can expand into with continuing growth also contribute to relatively constant population growth rates (McCullough 1990).

Bison in YNP seem to follow the cliff-edge model[1] (modified appropriately

---

[1]A model in which population growth proceeds at a constant rate until a point (carrying capacity) is reached abruptly, and growth rate drops precipitously to zero.

for constant recruitment instead of constant growth rate) but do not actually reach the cliff edge in that spatial limits of YNP were exceeded before carrying capacity was reached. It could be argued that the decline in annual increment in 1994-1995 and 1995-1996 (Figure II-6) reflected such a cliff edge. However, the decline was so clearly associated with hard winters, particularly in 1996-1997, that it seems more like a catastrophe than a reflection of exceeding carrying capacity on the basis of resource limitation. No equilibrium point is likely in a system in which the average annual population increment is 145 head, whereas once the population exceeds 3,000, the average SNOW condition results in an artificial removal of 332. That inequality is exacerbated by the unpredictable occurrence of mild and hard winters. It is to be expected that the population will build up until an inevitable winter reduction, only to repeat the process—much like kangaroo populations in Australia confronting periodic drought (Caughley 1987).

Meagher (1993) has noted that bison could be affecting their habitat in ways that will lower carrying capacity. YNP, however, states that no resource damage has been documented (Montana Department of Fish, Wildlife, and Parks et al. 1990). Taylor (1992) noted that bison killed outside the park in the hard winter of 1991-1992 were in excellent body condition, with more than adequate body-fat stores, and the same was true of animals removed in the hard winter of 1996-1997, which had substantial layers of brisket fat, particularly early in the winter (P. Gogan, USGS, pers. commun., 1997). Late-winter samples of brisket fat reflect access to forage rather than adequacy; therefore, early-winter values are more instructive than habitat quality. For the 1991-1992 kill, Zaugg et al. (1993) reported normal blood values and moderate parasite loads. Thus, there is little evidence of inadequate forage or quality available to YNP bison. Effects on habitat are a natural consequence of building populations, and no diminution of absolute population growth is apparent. Whether diminution of population growth will occur in the future can be determined only with time. Furthermore, periodic reductions of high populations by severe winters will reduce bison numbers and probably allow periodic recovery of the habitat. Given those uncertainties and the occurrence of a "natural experiment" because of the decline in the 1996-1997 winter, it is imperative that research be pursued.

Another issue is the effect of the 1988 fire that burned 42% of YNP. Boyce and Merrill (1996) modeled the expected ungulate-population response to the fire, and the model prediction was that bison would show a population increase. At least through 1994, no such response was reflected in the constant recruitment (Figure II-6).

Influences of Plowing and Grooming Snow

Meagher (1989, 1993) has described bison movement and suggested that learning of the landscape has gradually re-established behavior that was lost during periods when artificial control inside YNP confined bison to low numbers in a few areas of the park. Plowing and grooming of YNP roads in the winter for snowmobiles might have facilitated bison movements (Meagher 1989), and Meagher proposed that such pathways are energetically efficient to use, although there are no data to test that proposal. Limited plowing began in the late 1940s, but other than a few males, bison did not use the plowed roads until 1975-1976 (Meagher 1989). Grooming for snow-mobiles began in 1970, but according to Meagher (1993) the first bison use of groomed roads began in the winter of 1980-1981.

The fact that groomed roads were not used when first available (Meagher 1993) raises the question of why, if the groomed roads were valuable ener-getically and opened up valuable new habitat, bison did not quickly take advantage of the opportunity. The delay might be attributable to behavioral inertia. Or it might have been that bison had not yet reached numbers that forced expansion movement. Attributing bison population increase to road grooming instead of attributing use of groomed roads to population increase might therefore reverse cause and effect.

Still, the important issue is to separate behavior (proximate effect) from demographic consequences (ultimate effect). Expansion of bison into previ-ously unused habitat and movement through greater snow barriers would be expected simply because of the increase in the population (McCullough 1985). If grooming of roads led to substantial gains in winter energy savings (with presumed greater winter survival) and the opening of new habitat (with a presumed increase in carrying capacity), increased population growth would be expected. That proposition can be examined by looking at bison popula-tion growth before and after bison use of groomed trails began in the winter of 1980-1981. Absolute population growth rate was essentially constant before and after bison began to use groomed roads (Figure II-6); this finding suggests no substantial influence of snow grooming on demographic perfor-mance. There seems to be little supporting evidence of an ultimate effect of road grooming on bison population dynamics.

Bison were known to move along natural topographic routes before grooming began (Meagher 1973), and they cross barriers where roads do not occur at all (Meagher 1993; R. Garrott, Mont. State Univ., pers. commun., 1997). Furthermore, tallies of observations from Meagher (1993) showed 50 observations of bison using roads and 46 of bison traveling cross-country.

Those observations might have been biased in accordance with the observational effort in the two categories, but they show that cross-country travel by bison is common. Failure to prevent bison movements by hazing, herding, and fencing (Meagher 1989; Thorne et al. 1997) suggests that it will be difficult to prevent bison from moving where they please. Bison evolved in open plains largely as a nomadic species (Roe 1951; Meagher 1973). Just as other behavior—formation of large herds and other social behaviors—re-emerged with the increase in numbers after the bottleneck at the beginning of the 20th century, a nomadic tendency might be manifested in behavior of bison in YNP. The nomadic tendency is fostered by a large aggregate social structure in which individuals shift repeatedly between groups and the mother-calf relationship is the only consistent social bond (Lott and Minta 1983; Van Vuren 1983). Bison appear to behave as though continuous habitat were to be found down the valley or over the next hill, and that might account for their fluidity of movements when local conditions worsen. Whatever the case, now that locations of other habitat areas are known to the herd, it is unlikely that discontinuance of snow grooming will prevent their movements.

The suggestion that discontinuing winter road grooming will contain bison better within YNP and that starvation and other natural factors will relieve the need for artificial control outside the park appears optimistic. Certainly, periodic starvation of some bison in YNP during hard winters has occurred over many years (Meagher 1973). But many YNP bison in recent years have moved in search of better conditions elsewhere rather than attempt to survive winter in their traditional locale within park boundaries. The number of losses associated with movement out of YNP and being killed and the number of losses in the park can be examined. Of 1,805 total deaths listed by Meagher (1993) from 1975 to 1993, there were 1,127 outside and 678 (38%) inside YNP. In addition, natural mortality in the exceptionally hard winter of 1996-1997 can be estimated by subtracting the known kill in the winter of 1996-1997 and the 1997 summer count exclusive of new calves from the 1996 summer count. The summer count in 1996 was 3,436, of which 1,141 were killed or removed in the winter of 1996-1997, leaving 2,295 animals. If there were no natural mortality, 2, 295 adults should have been in the population in 1997; however, the highest summer count of adults was 1,921. The difference—374 animals (25% of the total mortality)—is the apparent natural mortality in YNP.

Gunther et al. (1997) reported bison and elk carcasses counted on systematic hiked, snowshoed, or skied routes (131.5 km) in various parts of YNP for 1992-1993 through 1996-1997. Mortality of bison varied from 5 to 22 from

1992-1993 to 1995-1996 and form a cluster around 20 per year, except for the low of 5 in 1993-1994, a low-SNOW (15.9 in.) winter. A total of 69 carcasses was found in the hard winter of 1996-1997, only one of which was in the low-elevation northern range. Bison carcasses are nearly significantly correlated with SNOW ($P = 0.07$), but that is entirely due to the leverage of the 1996-1997 point and is invalid because of violation of assumptions of the regression model. In addition, the current study of the Madison-Firehole bison showed that most of the deaths in the 1996-1997 winter were of calves, not adults, relatively few carcasses of which were found (R. Garrott, Mont. State Univ., pers. commun., 1997).

Those demographic analyses are subject to caveats. First, bison numbers are estimates and subject to all the errors associated with the problems of censusing wild animals in a heterogeneous habitat. But the population estimates given in Figures II-5 and II-6 show consistency over time that would not be expected if the error were large. The estimate is not likely to be so inaccurate as to invalidate the conclusion that most mortality occurred outside the park. Second, measurement error applies to weather variables as well, although these errors are more likely insignificant for the purposes of the analysis.

BISON IN GRAND TETON NATIONAL PARK
AND THE NATIONAL ELK REFUGE

Bison summer in GTNP and migrate to winter in the NER (Meagher et al. 1997). The history of this herd is given by Smith and Robbins (1994) and Williams et al. (1993). Twenty bison were reintroduced into Jackson Hole from YNP in 1948 and confined in the Jackson Hole Wildlife Park, a 1,500-acre enclosure for displaying prominent indigenous wildlife. A population of 15-30 bison was maintained in the park until 1963, when brucellosis was discovered in the herd. Several months later, all 13 adults in the population were destroyed, and four yearlings that had been vaccinated as calves and five newly vaccinated calves were retained. In 1964, 12 brucellosis-free adult bison (six of each sex) were introduced from Theodore Roosevelt National Park. Over the years, the enclosure's fence deteriorated, making it increasingly difficult to contain the bison. In 1969, when the remaining 16 captive bison in the herd were determined to be brucellosis-free, the herd was released to range freely. That event marked the beginning of the free-ranging Jackson bison herd. The current number is about 380 (S. Cain, GTNP, pers. commun., 1997), which shows that the population is continuing to grow. The herd is infected by brucellosis (76% seropositive, Smith and Robbins 1994; 36% culture-positive, Williams et al. 1993) and is in contact with infected elk on the winter feeding grounds in the NER, which are not being vaccinated. It is presumed that the brucellosis-free bison stock was originally infected on the NER feeding grounds through contact with aborted elk fetuses in about 1980 (Peterson et al. 1991b).

Bison and cattle have no contact on the winter range, because cattle are excluded from the NER. Bison are in contact with cattle as they cross private lands during migration, and cattle trail driveways in spring and fall and on grazing allotments on GTNP and Forest Service lands in summer (Smith and Robbins 1994). Cattle in this region of the GYA are invariably vaccinated because of the perceived risk of transmission of *B. abortus* from elk and bison to cattle. Smith and Robbins (1994) maintain that only one case of possible transmission of *B. abortus* from elk or bison to cattle has occurred in the GTNP-NER area since 1951, and it might have been due to incorrect vaccination rather than contact with wildlife.

ELK IN THE GREATER YELLOWSTONE AREA

As did bison, elk have consistently increased from low numbers since the beginning of the 20th century (see Figure 2). However, elk population size is substantially greater than that of bison. The current estimate of the number of elk in the northern range is about 17,000 (Figure II-7), and the total elk population in the GYA is around 120,000 (Toman et al. 1997). Traditionally, elk migrated out of YNP and GTNP to lower elevations, where, as their numbers increased, they became subject to public hunting under the authority of the wildlife agencies of the surrounding states. Elk are also extremely important to the socioeconomics of the GYA as a tourist attraction and game animal. Substantial hunter take of elk has generated less controversy than bison removals, presumably because of the much larger numbers of elk, their dispersion over a greater area, and more favorable perceptions of fair chase, and perhaps because the elk is less likely to be perceived as a national icon. Policy issues related to elk management have generated controversy, but more within the scientific and resource-management community than in the public arena. Because the debates over elk management are long standing, considerably more research has been done on elk than on bison. Herd units are recognized, and their population estimates are reported by Toman et al. (1997).

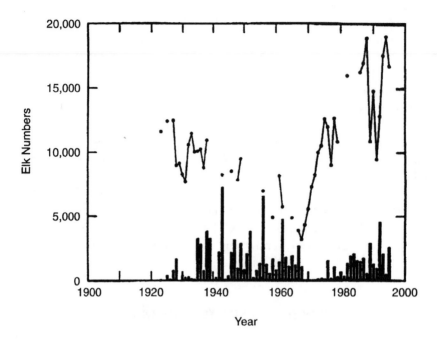

FIGURE II-7. Elk population estimates (circles) and hunting removals (bars) plotted on year for the northern YNP elk herd (data from Yellowstone National Park 1997).

Northern Elk Herd Movements
Out of Yellowstone National Park

In contrast with bison, there is substantial evidence that elk of the YNP northern range show density-dependent demography and are fluctuating about a dynamic equilibrium in response to resource carrying capacity, as well as being influenced by density-independent winter stress conditions (Houston 1982; Merrill and Boyce 1991; Coughenour and Singer 1996). The leveling of population growth since about 1980 is apparent from the plot of northern-herd numbers over time (Figure II-7). This herd was controlled by shooting in YNP until 1968; shooting in the park ended with the adoption of the natural-regulation policy. Elk kills thereafter (Figure II-7) were made during public hunting seasons on lands outside YNP under the control of Montana Department of Fish, Wildlife, and Parks.

The conclusion of other researchers about density-dependent control of the northern elk herd since the beginning of the natural-regulation policy in 1969 was rechecked with a somewhat different approach, including more-recent data. Change in elk population size between years (difference) was regressed against the elk population size in the first year, weather variables and indexes for the winter between the 2 years, and the elk kill in the fall and winter between the 2 years. Elk population size is the only variable that relates significantly to the difference in population size between years ($R^2 = 0.33$; $P = 0.03$) (Figure II-8). Regressions such as this have a small negative bias (Eberhardt 1970) because the Y-axis variable is not measured independently of the X-axis variable. Appropriate tests of density dependence have been much debated in the literature (White and Bartmann 1997), and no universally accepted method of analysis of non-independent data is available. Independent data, unfortunately, were not available to us. Still, the cessation of growth after 1987-1988 (apparent to the eye in Figure II-7), the relatively strong negative relationship in Figure II-8, and that various methods of analysis applied by others have led to the same conclusion of density dependence, are reason to accept tentatively that density dependence is more likely descriptive of the northern YNP elk population than density independence.

Including elk kill in more-complex models lowers the fit. Because the kills did not contribute to the prediction of the next year's elk population, either elk kills were swamped by density-dependent population changes, the kill was compensatory with natural mortality, or a combination of the two.

Although the dynamic equilibrium shown in Figure II-8 is characterized by high variance, the population size where difference = zero gives an estimate of mean carrying capacity. Note that if there is a negative bias (induced correlation) due to the analytical method, the mean will somewhat underestimate the true value. The negative bias is small, however, so this difference is probably negligible compared with the variance of the relationship. The dynamic equilibrium estimated elk population mean of about 11,300 includes the effects of elk removal by hunting. If kill is added back, the equilibrium population is about 17,400. Furthermore, the population estimates are corrected for sightability after 1986, and that results in higher estimates for 1986-1995 (Yellowstone National Park 1997). There is a high correlation between the estimated and corrected elk population size ($R^2 = 0.97$; $P < 0.001$). When elk difference plus kill is plotted against corrected elk populations (Figure II-9), an estimate of 17,812 is obtained, although the fit is not significant ($R^2 = 0.16$; $P = 0.20$). Despite the poor fit, this is a more realistic estimate of mean equilibrium and is in the range of values derived by other workers (17,058 by Houston 1982; 14,522-17,819 by Merrill and Boyce 1991;

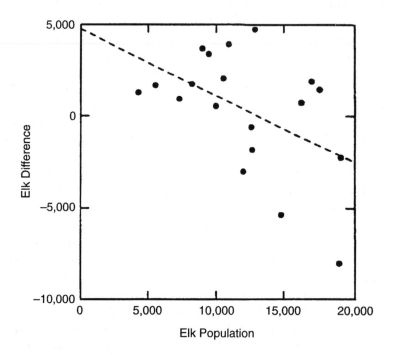

FIGURE II-8. Change in estimated elk population size between years plotted against elk population size in the first year for the years after 1969 when the natural regulation policy was implemented. The least-squares regression equation was $Y = 4731.951 - 0.362(X)$; $R^2 = 0.24$, $P = 0.04$.

18,010 by M. Taper (Mont. State Univ., and P. Gogan, USGS, pers. commun., 1997).

Natural regulation of elk population size occurs in the northern range of YNP (albeit with considerable amplitude in the dynamic equilibrium), in marked contrast with that of bison. Elk are mixed-diet feeders (Kufeld 1973; McCullough 1980; Singer and Norland 1996) and thus have a much higher amplitude in quality of diet. Taper (Mont. State Univ.) and Gogan (USGS, pers. commun., 1997) have shown that elk in the northern range follow a plateau and ramp model (McCullough 1990, 1992). That the effects of SNOW, hunter kill, and other variables seem to be dampened or compensatory with natural mortality suggests the risk of *B. abortus* transmission from elk to cattle is roughly stable in the northern Yellowstone range. And, of course, the seropositive rate in this elk herd is very low.

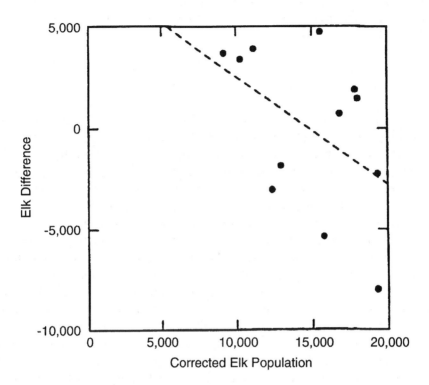

FIGURE II-9. As for Figure II-8 except using corrected elk population size. The least-squares regression equation was Y = 9333.435 - 0.524(X); R^2 = 0.16, P = 0.20.

Addressing issues of potential contact between elk and cattle requires identification of variables that influence the number of elk outside the park. Can a predictive model be found that explains elk movement, such as SNOW predicts bison movement at populations exceeding 3,000? The Montana Department of Fish, Wildlife, and Parks (unpublished data) conducted aerial counts from 1989 to 1997 to assess the number of elk leaving YNP in the northern range. Regression of these data on SNOW yields a significant relationship between snow water equivalent and elk migration out of YNP (Y = -3452.6 + 414.9(X); R^2 = 0.44; P = 0.05). Zero elk migration (on average) is at about 8 in. of SNOW; this suggests that elk are more easily moved by snow than are bison (17 in.). Furthermore, the kill of elk outside YNP in the Gardiner late hunt correlates with elk migrating out of YNP (R^2 = 0.67, P = 0.007). Carcass counts of elk in YNP (Gunther et al. 1997) correlate with snow

($R^2 = 0.83$; $P = 0.01$), and this also points to greater susceptibility to snowfall for elk than for bison. The equation predicts that no mortality will occur if SNOW is 15 in. or less.

Other Elk Herds in the GYA

Because of the controversy over the National Park Services's natural-regulation policy, the most attention paid to elk is paid to the northern-range herd in YNP, but that herd constitutes less than 20% of the elk in the GYA. One herd, the Madison-Firehole herd, is mainly (75%) nonmigratory and winters in thermal areas in YNP (Craighead et al. 1972). Six other recognized herds are scattered about the perimeter of the park; they typically summer within the boundaries of the park and winter outside it in three states—Montana, Idaho, and Wyoming. Each state has its own management program. Estimates of the size of these herds vary, but the total is around 120,000 elk. These populations are subjected to hunting outside YNP, and in a number of cases hunting mortality is great enough to limit the numbers. Migratory movements were curtailed historically by establishment of private ranches, and to maintain numbers and attract elk from private lands, winter feeding is common. In addition to the NER, the Wyoming Game and Fish Department (WGFD) maintains 22 feeding grounds (see Figure 2 for a map of wintering areas). Feeding in Wyoming usually occurs from January to April. Fencing and hazing are used to keep wintering elk confined to feeding grounds and separated from livestock. Idaho feeds elk on one area in years when it is necessary to keep elk off of developed lands (Smith et al. 1997). Montana does not provide prepared feed (alfalfa pellets) for elk but has made an alfalfa planting available to wintering elk at Dome Mountain (see numbers using this area in Coughenour and Singer 1996).

In addition to the roughly 23,000 elk on feeding grounds (GTNP 1993), about 25% of the Wyoming elk in the GYA winter on natural range away from livestock operations. The WGFD has embarked on a program to improve habitat in natural wintering areas to reduce the density of elk on the feeding grounds and disperse elk over the landscape (S. Smith, WGFD, pers. commun., 1997). Elk have been found to use areas improved by controlled burning. However, it is not yet clear whether there is sufficient opportunity to spread the population adequately or to reduce substantially the numbers of animals using the feeding grounds. Oldemeyer et al. (1990) have shown that weight loss in elk on the NER is related to the amount of feed given and that elk use natural feed if it is available. Thus, density on feeding grounds potentially could be reduced by a combination of greater hunter harvests, manipulating the amount of feed offered and its placement, and making natural feed more available. Many Wyoming elk herds are above herd objectives (Toman et al. 1997), and increased hunter harvests are desired to bring numbers into compliance with management goals. This would reduce the need for supplemental feeding and the consequent crowding of elk where transmission of *B. abortus* is most probable.

Presumably, reduction of density on the feeding grounds would reduce the likelihood that elk would come into contact with infective products of abortion due to brucellosis and would reduce the rate of transmission of *B. abortus*. Whether those measures will be sufficient to reduce the incidence of brucellosis in elk remains to be seen. It seems likely that if females abort away from the feeding grounds, the rate of transmission will be reduced, leading to a reduction in the overall herd infection rate. Nevertheless, it will be difficult to reduce elk density on the feeding grounds enough to prevent transmission from abortions and avoid maintaining a problematic level of infection.

There is a small interchange of individuals between these herd units (Anderson 1958; Craighead et al. 1972; Boyce 1989), but it is only a few percent. As noted earlier, the cause of seropositivity in the northern herd is an issue. Although the exchange of individuals between herd areas, including exchange between Jackson Hole and the northern range (Craighead et al. 1972), shows that the seropositive elk could have come from southern herds to winter with the northern herd, it does not show that they did. Yet, interpretation of the presence of seropositive elk in the northern range depends entirely on whether these are the same animals that came from herd areas with high seropositive rates, rather than animals that came into contact with *B. abortus* through contact with other elk or bison.

The best-studied population other than the northern herd is that wintering on the NER and vicinity (Anderson 1958; Boyce 1989; Smith and Robbins 1994). In his extensive population analysis, Boyce (1989) concluded that NER elk show density dependence despite winter feeding and that they are regulated by winter severity, recruitment, and hunter harvest. Hunter harvest appears to have a larger effect on NER elk than on the northern YNP elk, but the harvest rate is also higher. Boyce (1989) reported an annual kill over many years of around 3,000-4,000 in a population of about 10,000-12,000 (25-30%). He estimated maximal sustainable yield to be about 30%, which suggests that this population has been harvested fairly near its maximal rate. Some agency biologists suggest that this estimate of maximum sustainable yield might be high due to underestimation of the herd size and inclusion of harvest from other herd units. From feedground classification counts, they suggest that maximum sustainable yield is nearer 20%. In either case, the harvest rate on this population is relatively high. In the northern herd, by comparison, the annual kill in recent years has been about 1,800 in a population of 16,000 (11%) (Yellowstone National Park 1997).

It is clear that in contrast with the northern herd, which is limited mainly by natural phenomena, the other herds using the GYA are limited mainly by human harvest. Thus, they are more stable from year to year in their likelihood of contact with cattle and with the consequent possibility of transmission of *B. abortus*. Essentially, they come out of the YNP area, pass through a hunting zone, and are intercepted by winter feeding areas. To the extent that feeding areas do not stop movement, elk are hazed to return to them from private lands. Although the other herds are more predictable from year to year, the sheer numbers of elk, their proximity to grazing allotments, cattle trailing areas, and private ranches, and their relatively higher seropositive rates means that the relative risk of transmission of *B. abortus* from elk to cattle is greater than for the northern herd elk.

EFFECTS ON REPRODUCTIVE POTENTIAL

Two questions arise when considering whether *B. abortus* affects the reproductive potential of bison generally, and specifically bison in the GYA. The first question is whether *B. abortus* lowers the reproductive rate. That question would be consistent with the traditional use of the term *potential* in wildlife management, in which it is viewed as the maximal possible rate of reproduction (Leopold 1933). By that definition, the answer to the question is yes because *any* abortion or lowering of the probability of survival of offspring—the usual manifestation of brucellosis in bison, as in cattle and elk—would reduce the maximum. However, such a strict definition probably is not the most relevant in the context of brucellosis in bison in the GYA.

The second question deals with whether brucellosis affects the population dynamics of bison, and this is more relevant to the current issues in GYA. The question is whether brucellosis lowers reproductive performance sufficiently to constitute an important factor in population dynamics in bison and thereby alters the population trend over time.

Controlled research on the magnitude of brucellosis effects is lacking, but it can be estimated from the modeling results of Peterson et al. (1991b). They modeled bison populations (females only) under brucellosis-free and brucellosis-infected states. Their projections for a brucellosis-free population can be used to estimate the impact of brucellosis on the growth rate of the GTNP bison population (69% seropositive for *B. abortus*) from the 1970 escape of five female founders to the total 1989 female population. This result can be derived from their Figure 7 (panel B). The annual growth rate projected by their model was 15.48%, whereas the realized rate was 14.45%, 1.03 percentage points lower. Simple models that assume infection rates between about 10% (GYA culture-positive rate) and about 50% (GYA seropositive rate) and loss of the first calf after infection show that reduction of population growth rate because of brucellosis would be only a few percent unless the survivorship of reproducing females were extremely low, an unlikely possibility for the hardy, long-lived bison.

Empirical results bear out that conclusion. Bison populations in YNP and GTNP (and herds elsewhere) have continued to increase despite being infected with *B. abortus* (Figure II-2) unless artificially controlled or reduced by severe winter conditions (Dobson and Meagher 1996). In YNP, artificial removal has been important in holding bison population growth to near zero at times, particularly from 1935 to 1965, when the herd was managed to number around 400, and in the past few years (Figure II-2). Among natural variables, winter mortality is clearly the most important, but production of forage in summer also might contribute to the dynamics of the herd (Meagher 1973). Only for a bison population in a marginal habitat where it would be barely capable of holding its own would brucellosis be the deciding factor in survival of the herd. YNP is not marginal habitat.

Elk, like bison, will suffer decline in potential elk population growth due to abortion. Although the data for elk show greater variance than those for bison, the persistent increase in numbers of elk after declines have resulted in brucellosis being considered unimportant as practical matter.

RISK OF TRANSMISSION

The risk of transmission is determined largely by the number of abortions that occur, the presence and survival of *B. abortus* in placental exudates, and the exposure of a susceptible host through an appropriate tissue barrier. Aborted placentae might contain as many as 10^{13} *B. abortus* per gram of tissue (Davis et al. 1995). Direct evidence of transmission from various wildlife species to cattle has been difficult to establish. Despite circumstantial and epidemiologic evidence of transmission, many still believe that bovine brucellosis never has been proved to be linked to free-ranging elk or bison. The detection of transmission of *B. abortus* from an infected animal to a susceptible domestic cow is complicated by lack of clinical signs in infected cattle, geography, predation of the placenta and fetus, and birthing character-istics of the infected animal. Those and other factors complicate determina-tion of risk of transmission. The perception of, for example, some members of animal-welfare groups, is that transmission is extremely rare and might never occur. The perception of others, such as some ranchers in or near the GYA, is that seroreactive cattle do appear in their herds and that those cattle have been infected with *B. abortus* from either elk or bison.

Bison to Cattle

Under natural conditions, the risk of transmission from bison to cattle is very low, but the appropriate quantitative risk assessments have not been done; one, by a multiagency group, is under way (E. Williams, U. Wyom., pers. commun., 1997). Free-ranging bison or elk might have served as the source of *B. abortus* infection in six cattle herds in the GYA (GYIBC 1997), but as noted earlier, the evidence is ambigous. Transmission of brucellosis from naturally infected captive bison to cattle has been reported; captive bison under range conditions in North Dakota were in contact with beef cattle during the winter (Flagg 1983). Bison-to-cattle transmission in Arkansas has also been reported.

The risk of transmission of *B. abortus* from infected bison to cattle is a major part of this study. Brucellosis has been transmitted from bison to cattle under experimental conditions, and brucellae were transmitted from infected bison to seronegative cattle when the animals were confined to-gether in pens (Davis et al. 1990).

YNP bison herds have had little or no contact with outside bison since the early 1900s. Serologic surveys show seroprevalence rates of 20-73% (Rush 1932; Tunnicliff and Marsh 1935; Clark and Kopec 1985; Pac and Frey 1991; Aune and Schladweiler 1992; Aune et al. 1997). The number of abortions or fetal deaths per 100,000 bison births since brucellosis was first detected in 1917 is not known, and individual cases of transmission, especially in early periods, will likely never be determined. In the past decade, two cases of abortions due to *B. abortus* have been established (Rhyan et al. 1994).

Isolates of *B. abortus* obtained from bison have been shown to be patho-genic in cattle; for example, biovar 1 isolates from a Wood Buffalo National Park bison in Canada were virulent when inoculated into cattle (Forbes et al. 1996), even though the bison had been segregated from cattle for more than 60 years.

The current risk of transmission from YNP bison to cattle is low. Further-

more, domestic cattle adjacent to the park are vaccinated, cattle are monitored by federal agencies, and ranchers are vigilant.

Elk to Cattle

Transmission of *B. abortus* from elk to cattle is unlikely in a natural setting.
The ability of brucellae to be transmitted from elk to cattle under experimental conditions has been proved (Thorne et al. 1979), however, and cattle
mingling with aborting elk on feeding grounds would be at high risk for
infection. Elk densities in YNP reach those of the winter feeding grounds (p.
76, "Other Elk Herds in the GYA") for short periods during some times of the
year; although the incidence of brucellosis in these elk is very low, that might
present another risk factor. Data on the incidence of elk-to-cattle transmissions might be skewed if ranchers are not forthright in admitting when cattle
might have been exposed by commingling with infected elk.

Elk to Bison

Elk can transmit *B. abortus* to bison. Transmission is probably limited to
aborting and parturition of infected elk with release of fetal membranes and
genital exudates that contain large numbers of *B. abortus*. That has occurred
during mixing of bison with infected elk on feeding grounds of the National
Elk Refuge. M. Meyer (U. Calif., pers. commun., 1997) claims that the Jackson
(GTNP) bison herd was brucellosis-free until it discovered the elk feed lines.
The Jackson herd, which for 20 years was confined in a wildlife park and
allegedly was brucellosis-free, escaped in 1968 and commingled with infected
feeding-ground elk around 1980. The herd became infected (the seroprevalence in 35 bison collected in 1989-1990 was 77%) either by elk on the National Elk Refuge (NER) or by bison that were infected (although seronegative)
when they escaped.

Transmission from elk to bison might have occurred under natural conditions in the GYA (Williams et al. 1993). If low infection rates are attained
through management of bison, the population of bison will remain uninfected
for quite some time before a low-probability elk-bison or bison-elk transmission would occur. During the brucellosis-eradication program in Custer State
Park in South Dakota, elk, deer, and antelope mingled with infected bison;
there is no evidence that bison from which brucellosis was eliminated were
reinfected by *B. abortus* from elk (Gilsdorf 1997).

Bison to Elk

Whether and at what rate *B. abortus* is transmitted from bison to elk are unknown (S. Olsen, USDA, pers. commun., 1997). One group states that "although controlled or field studies have not been done to establish transmission between bison and elk, it certainly is possible" (T. Kreeger, Wyo. Game and Fish, pers. commun., 1997). Evidence of transmission of brucellosis among wildlife species comes from Elk Island, a fenced national park in Alberta, Canada, where bison were believed to have been the source of *B. abortus* infection in elk and probably moose (Corner and Connell 1958).

Elk as a Reinfection Pathway for Bison

Bison can contract *B. abortus* from elk, as demonstrated by the case cited above in which a clean herd of bison was introduced in 1970 to GTNP, later wintered on the elk feeding grounds of the NER, and tested positive for *B. abortus* in 1989. The risk of transmission to bison will depend on the success of efforts to reduce the infection rate in elk by vaccinating elk on feeding grounds and dispersing them over a larger wintering area in the southern GYA. If infection rates are not substantially reduced in elk, it seems inevitable that reinfection of bison will occur, just as bison are a continuing reinfection source for elk (Thorne et al. 1997). It must be remembered that low-probability events multiplied by large-enough animal contacts over a long-enough time become inevitable events. Apparent multiple transmissions between some combination of cattle, bison, and elk with the arrival of *B. abortus* in the GYA (Meagher and Meyer 1994; Thorne et al. 1997) should be a cautionary note, as should the occurrence of a case of undulant fever in an elk hunter in the northern range (where seropositive rates of elk are low), whereas no cases in hunters in the southern range (where seropositive rates are high) are known.

Early work with vaccination of elk at Greys River winter feeding ground resulted in promising reductions in seropositive rates (67% to 12%). In the winter of 1996-1997, however, the rate rebounded to 26%. The cause of the increase is unknown, but it could be related to the hard winter of 1996-1997, which would indicate that environmental stress, as well as pregnancy stress, can contribute to *B. abortus* infection rates. The reduction from 67% (1976) to 12% (1996) is significant, but inclusion of the 1996-1997 point results in lack of significance (P = 0.11). It is problematic that the first year in the time series (1976) was 17 years before the first of the consecutive data points

(1993-1997). Also, Smith and Roffe (1997) have questioned the validity of conclusions from the vaccination experiments on which the program is based. Alternatively, the field-vaccination program might be reaching the limits of its efficiency. For example, modeling of vaccination in bison (Peterson et al. 1991a) and cattle shows that reduction in seropositive rates amount to only 60-80% of baseline prevalence. That would predict that elk vaccinated on the feeding grounds would show a reduction in seroprevalence from 67% to about 13-27% , the approximate range observed in recent years. Further work will be necessary to evaluate the success of the program.

The source of the 1-2% seropositive rate in elk in the YNP northern range is potentially important. That seroprevalence might be, as has been proposed, the result of movement of elk from southern to northern ranges. But alternative explanations need to be considered, such as the infection of northern-range elk by contact with infected southern-range elk or YNP bison. The calving areas of southern-range elk, where birth and abortion increase the likelihood of transmission, are well south of YNP (Boyce 1989), and this casts doubt on that source of infection. If movement of southern-range elk to the northern range is not responsible for the seropositive rate in northern elk and if northern-range elk are not in close contact with southern-range elk, then the rate would seem to be natural infection due to contact of elk with infected bison. Because the potential of such transmission—either between bison and elk or between elk away from the winter feeding grounds, a key issue in sustainability of *B. abortus* in non-feeding-ground elk—is of particular interest, it is important to determine whether the seropositive elk in the northern range have moved from the southern range by marking them on the feeding grounds through feed or vaccination.

Several factors contribute to the likelihood of potentially infective contact of bison with elk. First, the distributions of the two species overlap broadly in the GYA on the summer range, where they are more dispersed, and on the winter range, where they are concentrated (Meagher 1973). Bison and elk are often seen near each other.

Second, their habitat requirements overlap broadly. In YNP, Singer and Norland (1996) found overlap of diet (1 = complete overlap, 0 = no overlap) to be 0.47 and 0.63, and use of vegetation 0.43 and 0.75, slope/aspect 0.45 and 0.57, and snow 0.59 and 0.89 for early and late periods. Those overlap values are high, and that they invariably were higher in the later years suggests that increases in density in both species are increasing the overlap of their use of the environment. Similar overlap between bison and elk was reported by McCullough (1980) for the National Bison Range in northwestern Montana and by Telfer and Cairns (1979) for Elk Island National Park, Alberta.

Thus, increases in potential contacts are not a simple function of numbers, but a function of the increased forcing of overlap of niche space as well.

Third, the probability of transmission of *B. abortus* needs to take into account a behavioral component. The movements of the two species are essentially independent. If encountering a site (such as a birth or abortion site) at which *B. abortus* might occur in the environment were random, the probability of transmission would be low by chance. However, birth and abortion sites are likely to attract both species. Bison and elk are highly olfactorily oriented. The observation of W. Cook (U. Wyoming, pers. commun., 1997) of attraction of elk and bison to noninfective bovine fetuses placed in the environment illustrates the point. Such attraction might occur across species as well. Reproductive fitness is a major component of natural selection (Fisher 1930), so it behooves individuals to be cognizant of each other's reproductive state. Bison—especially males—show substantial interest in matters associated with reproduction. Berger and Cunningham (1994) discuss these aspects in detail for bison. Elk and bison engage in *flehman*, a behavior most commonly associated with males testing the estrus state of females by licking the vulva or urine and exposing the molecules therein to the vomeronasal organ in the palate. Bison males have been reported as displaying "very aggressive behavior towards cows in estrus, or any blood discharge, death or injury" (S. Holland, as cited by Kearley 1996). They are especially animated by the occurrence of aseasonal estrus (S. Holland, state vet., S.D., pers. commun., 1997) or by blood at any time (J. Rhyan, APHIS, pers. commun., 1997), and the aseasonality of abortion might evoke similar interest.

Behavioral attraction to sites of abortion or birth, therefore, is likely to bring individuals into contact with potentially infective materials at a rate far greater than expected from random movements. That behavior is most prevalent in bulls, but it occurs to some extent in cows. That would increase the probability that bison, especially males with greater movements and sexual curiosity, would be infected by *B. abortus* shed by elk. In fact, in the bison herd observed by Holland, bulls were more likely than cows to become infected by cows. That also is true of the GYA. Bulls tested in the winter-killed sample leaving YNP had a 57% culture-positive rate compared with 24% for cows; for GTNP, bulls had an 84% seropositive rate compared with 69% for cows. Furthermore, the difference is already present in subadult males (Meyer and Meagher 1997).

The higher prevalence of brucellosis in bison bulls than cows is puzzling. The difference might arise from differential survival of offspring by sex. If birth to infected mothers or acquisition of *B. abortus* through the milk (see p.

23, "Shedding in Mammary Glands and Milk") were the source of infection, calves of both sexes would have similar infection rates unless there were differential abortion or calf mortality by sex. We assume that mortality is more likely among infected fetuses or calves than among uninfected ones. Abortion would have to be more prevalent for female than male fetuses to account for the differences. Ordinarily the reverse would be expected—the male is larger and places greater stress on the mother. Furthermore, abortion is thought to be relatively uncommon in YNP bison, so differential abortion by sex is not an expected source of the different infection rates. Differential mortality of male calves is not likely the cause, in that uninfected calves would have to have a higher mortality. The difference could be due to differential mortality of infected female calves, but this would invert the logic in explaining why female infective rates were low rather than why male rates were high. Although possible, this explanation seems at odds with the typical higher male calf mortality that results in a prevalence of females among adult bison. Higher seropositive rates in bison males are unlikely to be due to differential mortality by sex in calves.

A more likely possibility is that higher infection rates in males arise from differential behavior of males later in life. Male behavior that might contribute to infection includes naso-oral contact with genital exudates and urine during flehman to ascertain estrus, greater inclination to smell or lick afterbirth or aborted materials (even subadult males would be subject to this route of infection), and venereal transmission during coitus or contact in tending bonds, which because of the polygynous breeding behavior of bison brings each male into contact with multiple females.

It is notable that elk show the reverse condition: females have higher seropositive rates than males. Ordinarily, elk bulls are spatially segregated from females except during rut (McCullough 1969; Geist 1982). Like bison, elk bulls perform flehman during rut and mate with multiple females. They do not form tending bonds but instead guard harems; this lessens the period of close contact and reduces the number of males involved in copulation. For example, McCullough (1969) found that only 12% of male elk were important contributors to reproduction, whereas in Badlands National Park, 51% of 37 bison males 4-9 years old mated (J. Berger, U. Nev., pers. commun., 1997). In general, elk seem less alert to strange odors outside of rut than do bison. That means that male elk on the winter feeding grounds are less likely to be curious about abortions in the winter time than are females. All those factors suggest that bison bulls are more instrumental in transmission of *B. abortus* than are elk bulls. As noted earlier, the role of bison bulls in transmission of *B. abortus* presents an important gap in our knowledge.

Other GYA Wildlife to Cattle

Infection with *B. abortus* is self-limiting in many wild mammals. Brucellosis occurs rarely in deer, pronghorn antelope, and mountain sheep. Brucellosis has not been documented in those species in the GYA, and any infection in them would be inconsequential for the control of brucellosis in bison and elk populations. Natural infection with *B. abortus* in avian species has been reported (Angus et al. 1971) but plays no role in transmission to mammals.

Transmission to Humans in the GYA

Human infection with *B. abortus* in the GYA has been reported, and a woman elsewhere was reported as having aborted due to *Brucella* spp. Hunters consume bison meat from areas outside YNP, and some bison meat is given to tribal peoples and soup kitchens for needy people. Meyer reports that "in December 1991-February 1992 over 500 bison were shot and carcasses were eviscerated, largely by Indians who literally just mucked through the guts" (M. Meyer, pers. commun., 1997). However, no evidence that *B. abortus* infected those Indian populations has been reported.

The U.S. Centers for Disease Control and Prevention no longer requires reporting of undulant fever. The World Health Organization *Laboratory Biosafety Manual* places *B. abortus* in risk group III, indicating a high risk to persons involved in handling infected animals or tissues. All personnel involved in sampling should be formally advised of the risk of infection and trained in the handling of infectious tissue and the use of masks and equipment. Face masks, gloves, and protective clothing should be used in high-risk situations that involve female bison that have placental lesions of brucellosis or that have aborted.

The greatest risk of human infection lies in body contact with infectious material and transmission of microorganisms from hands to body orifices. *B. abortus* is typically present in low numbers in blood and lymphoid tissues of animals. Although most genital tissues in males and nonpregnant females have only low numbers of organisms, the infected placenta and its fluids are extremely hazardous and can contain up to 10^{13} bacteria per gram, a concentration that makes aerosol transmission possible. Blood and milk are hazardous, but infection from them is unlikely if reasonable means are used to prevent contamination of hands and face and thereby controlling the potential to spread microorganisms to body orifices. Pasteurization of milk eliminates the risk of infection from milk consumption.

Human brucellosis caused by *B. suis* has been well documented in people who eat rangiferine animals (reindeer and caribou) in North America (Ferguson 1997), but no systematic study of brucellosis in American Indians has been done. Brucellosis was identified in Eskimos in Canada (Toshach 1963). In Alaska, 49 cases have been reported. It was suggested that rangiferine brucellosis is widely underreported because mild cases are not brought to medical attention and that chronic human cases might be undetected by small medical clinics.

Treatment of human brucellosis involves 4-6 weeks of antibiotic therapy, which carries the possibility of toxicity in some patients. Cure is not ensured, especially in chronic disease, which can be lifelong.

OTHER SPECIES OF *BRUCELLA* AND BRUCELLOSIS IN WILDLIFE

Species of *Brucella* other than *B. abortus* are associated with brucellosis in wildlife (Table II-1). Rangiferine brucellosis in commercial herds of reindeer

TABLE II-1. Species of Brucella			
Bacterium	Primary Hosts	Wildlife Hosts	Pathogenicity in Humans
B. abortus	Cow	Bison, elk, wolf, coyote	+ +
B. melitensis	Goat, sheep	Camel, wild ruminants	+ + +
B. suis	Biogroups 1 and 3: Pig Biogroup 2: Pig and hare Biogroup 4: Reindeer and caribou Biogroup 5: Rodents	Reindeer, pig, caribou	+ + +
B. ovis	Sheep	Mountain goat	-
B. canis	Dog	?	+
B. neotomae	Woodrat	Desert woodrat	-
B. sp (unnamed)	Dolphin, seal	Many marine mammals	+

and in caribou (*Rangifer tarandus*) throughout North America is caused by *B. suis* biovar 4. *B. suis* in these animals has been shown to infect cattle. In one study, four of eight cattle penned with 14 naturally infected reindeer became infected with *B. suis* and were seropositive (Forbes and Tesaro 1993). Although it has not been reported, reindeer likely can transmit brucellosis to other wild mammals, and this could cause confusing serologic responses in bison (Forbes and Tessaro 1993).

A variety of rodents have been reported to be susceptible to infection and to develop disease (Moore and Schnurrenberger 1981), but rodents have not been implicated in the spread of brucellosis in the GYA, perhaps because of inadequate investigation.

III
VACCINES

Bison in captive herds usually are vaccinated using protocols established for cattle, and some elk are vaccinated at feeding grounds. Vaccination in bison and elk is one part of an overall strategy that could be used to control or eliminate *B. abortus* in the Greater Yellowstone Area (GYA), but much research is needed before current vaccines can be judged adequate for use in those species. The following sections discuss current vaccines and describe the biosafety and efficacy standards that new and existing vaccines must meet if they are to be used effectively as part of a control or eradication program.

EXISTING VACCINES

Four vaccines are used against brucellosis: *Brucella abortus* strain 19 (S19) and *B. abortus* strain RB51, Rev 1 against *B. melitensis*, and strain 2 against *B. suis*. S19 and RB51 were developed to prevent brucellosis in cattle and have been used in bison and elk. They do not protect completely against infection or abortion in cattle, and data on their limited use in bison and elk suggest a similar degree of protection. Although modern techniques of molecular biology have revealed differences in the arrangement of DNA sequences between these vaccine strains and the virulent field strains, the differences are not reflected in the antibody responses of the host.

Further research on the vaccines is required before adequate data will be available for interpretation of safety and efficacy. If the vaccines prove inadequate, several approaches are available for vaccine development: development of a new or mutant strain of *B. abortus* designed for use in bison and exploration of the use of adjuvants in association with vaccines. Yellowstone National Park (YNP) bison are not needed for human consumption so some adjuvants not approved for cattle might be useful in bison and elk.

Strain 19

B. abortus strain 19 is a low-virulence, live vaccine developed for use in cattle. It is cleared from the body of the cow more quickly than is the virulent field strain from which it was derived. S19 lacks the *eri* gene for erythritol metabolism, but it is uncertain whether that gene is associated with virulence. S19 was the cornerstone of the United States Department of Agriculture program of brucellosis eradication in cattle from the 1930s to 1996. When given to cattle in calfhood, S19 has been shown to be only 67% effective in preventing infection and abortion. It has several disadvantages: it is infectious for and causes disease in humans; when given to pregnant cattle, it infects the placenta and can cause abortion; and it induces serologic responses in vaccinated calves that cannot be discriminated from serologic responses caused by field infections.

S19 in Bison

In commercial bison herds, S19 has been used for calfhood vaccination since the 1960s without important clinical sequelae. Commercial producers using S19 in bison typically use vaccine doses established for cattle. The standard dose of S19 for cattle was originally required to contain at least 10 billion live cells per milliliter (50-billion dose) on initial test and at least 5 billion per milliliter (25-billion dose) at expiration date. A reduced dose of S19 was later established as 0.3-3 billion live cells with age limits of 4-12 months for use. The reduced dose range was based on data for minimal protective doses of 0.09-4.5 billion colony-forming units (CFU) for calves 3-6 months old (Davies et al. 1980) and 0.1-90 billion for calves 4-6 months old (Deyoe 1980).

In South Dakota, which has more bison than any other state, many of the 200 small commercial bison herds are vaccinated according to state regulations. Bison are moved through chutes and vaccinated subcutaneously. One commercial herd of some 5,000 bison in South Dakota (Triple U) was vaccinated with S19 beginning in the 1960s and with RB51 since 1996.

Attempts were made in the 1940s to control brucellosis in YNP bison. A vaccination program with S19 was begun and was believed to have achieved some success in reducing the incidence of brucellosis (Barmore 1968).

Records of that program are not available, and it is not possible to judge the effect of the program on YNP bison.

In bison herds in national and state parks during the 1960s, calfhood vaccination with S19 was part of programs to control brucellosis. The herds were chronically infected with *B. abortus,* and within a decade of the start of an S19 vaccination and management (including test and slaughter) program, many of the herds were declared free of brucellosis (see Chapter IV). In most cases of S19 use in bison, no abortion, anaphylaxis, lameness, or other sequelae associated with vaccine use have been noted. However, records of vaccine use and effect have not been kept for most herds.

Results of biosafety and efficacy studies of the use of S19 in adult and pregnant bison suggest caution. In some experiments, S19 appeared to be more virulent in bison than in cattle and caused a high incidence of abortion when given to pregnant bison.

Experiments were done with 18 pregnant bison given 1×10^7 CFU of *B. abortus* strain 2308 in the conjunctival sac. In one experiment, 12 infected bison were placed in contact with 12 susceptible pregnant heifers: six aborted and two had nonviable calves, compared with nine abortions in 12 similarly inoculated cattle. Five of the 12 susceptible cattle became infected (Davis et al. 1990). In a second experiment designed to define pathogenesis, six infected bison were killed, one each week. All tissues were negative at 1 wk, only parotid and supramammary lymph nodes were positive at 2 wk, and most lymph-node samples were positive from 3-6 wk. Thymus was never positive, and spleen was positive only at the sixth week.

At a reduced dose in adult bison, S19 does induce some protection against experimental challenge but also induces a high percentage of abortions (Davis et al. 1991). When adult female bison were vaccinated with S19, abortion and infection rates were reduced in comparison with nonvaccinated bison; however, S19 caused pregnant bison to abort (Davis et al. 1991). S19 vaccine was not effective when administered to (bison) calves (Davis, unpublished data, quoted in Thorne and Herriges 1992). Abortion in bison experimentally vaccinated with S19 has been well documented, and *B. abortus* has been shown to replicate in epithelia of the trophoblast as it does in cattle and to cause placental inflammation and necrosis, changes that underlie abortion (Davis et al. 1990).

The responses of bison to S19 might differ from those of cattle for several reasons, including increased susceptibility of the host, vaccination of the host at an inappropriate age, and differences in social behavior that favor transmission of *B. abortus* (F. Enright, LSU, pers. commun., 1997).

S19 in Elk

Nearly 36,000 doses of S19 are estimated to have been given to elk, and the vaccine is considered safe and efficacious. Elk have been vaccinated on Wyoming feeding grounds with a reduced dose of *B. abortus* S19 given by biobullet (Thorne et al. 1981; Herriges et al. 1991), and a significant decrease in serum antibody titers has been reported (Smith et al. 1995). The Wyoming Game and Fish Department (WGFD) began giving S19 vaccine to elk on WGFD-managed feeding grounds in 1985. More than 35,000 doses of vaccines have been given on 21 elk feeding grounds. At one feeding ground, the prevalence of brucellosis has declined by 50% (P = 0.00001). The decline is attributed to vaccination because other management practices have not changed over 25 yr. No environmental hazard has been associated with S19 vaccine use for elk in Wyoming.

Although elk at Jackson Hole have been vaccinated with S19, serologic studies without culture studies will not give a true picture of the prevalence of brucellosis in the elk population, because the serologic tests cannot differentiate between a titer caused by the vaccine strain and that caused by field strains (D. Ewalt, Nat. Vet. Serv. Lab., pers. commun., 1997).

Strain RB51

B. abortus strain RB51 is a rough mutant of virulent *B. abortus* strain 2308 that is deficient in O-side chains of lipopolysaccharides on the bacterial surface. RB51 was naturally derived by serial passages on media containing rifampin and by selecting single colonies with rough morphology. The genome of RB51 closely resembles the genome of strain 2308 when examined with most molecular techniques. However, RB51 has a unique genetic rearrangement that differentiates it from strain 2308—one that is stable and has been maintained in all isolates of RB51 (Figure III-1). Genomic restriction-endonuclease patterns produced with pulsed-field gel electrophoresis have also demonstrated a unique "fingerprint" for RB51 relative to other brucellae (Jensen et al. 1995, 1996). After passage in vitro or in vivo, RB51 retains its resistance to rifampin or penicillin and its susceptibility to tetracycline. A new growth medium has been developed for culture of RB51 (R. Hornsby, Nat. Anim. Dis. Ctr., pers. commun., 1997).

RB51 has replaced S19 as the required vaccine for cattle in the United States. It is genetically stable in bison, as it is in cattle, and does not revert to virulence or to smooth forms after growth in vivo. The dose used commer-

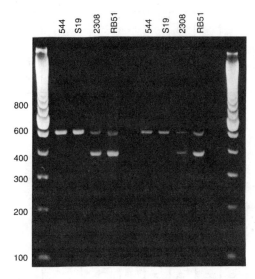

FIGURE III-1. Differentiation of *B. abortus* vaccine strains S19 and RB51 by a polymerase chain reaction assay (Bricker and Halling 1995). PCR amplification was done on DNA from 4 bacterial strains. DNA was resolved by electrophoresis in agarose, stained, and photographed. Sources of DNA (top) and sizes (in base pairs) of fragments of the 100-bp marker in lane 1 (left).

cially is 10-34 billion live organisms delivered in 2 mL. At that dose, RB51 has been shown to be protective in cattle when used as a calfhood vaccine between 3 and 10 months of age (Cheville et al. 1996). RB51 studies in bison and elk are shown in Table III-1 (Olsen et al. 1998). It is noteworthy that vaginal, rectal, ocular, and nasal swabs collected from bison 1-18 wk after vaccination in the second Olsen study listed in the table did not contain culturable *B. abortus*. Numbers of bison that were culture-positive in prescapular lymph node biopsy at 6, 12, 18, and 24 wk after infection were 4 of 4, 3 of 4, 1 of 4, and 0 of 3, respectively. Quantitative data suggested that most bison will clear RB51 within 18-24 wk.

RB51 has tropism for the bison placenta. It has been shown experimentally to cause endometritis and placentitis that result in abortion in pregnant bison. Of eight bison cows given RB51, two aborted (68 and 107 days after vaccination). RB51 was isolated from the cows' reproductive tissue and supramammary lymph nodes and from fetal bronchial lymph nodes and amniotic fluid. The vaccine dose used was similar to that used in cattle and might not be appropriate for bison (Palmer et al. 1996).

RB51 might not be as effective in bison and elk as in cattle, although data on this are not adequate to know. Studies of other species have yielded variable results. Experiments on the efficacy of RB51 against *B. ovis* in rams showed no protection (Jimenez de Bagues et al. 1995). RB51 was given at 4 x 10^{10} CFU subcutaneously; 6 months after vaccination, rams were challenged with 3 x 10^9 CFU of *B. ovis* and examined 8 wk later. Controls and RB51 vaccinates were found to be 100% infected, compared with 68% infection after use of Rev 1 vaccine in rams.

TABLE III-1 RB51 Vaccine Biosafety Tests

Author	n	Animals	Route	Dose	Site	Study
Elk						
Elzer, Davis	19	adult	oral	$2\text{-}3 \times 10^{10}$	North Dakota	vaccination challenge
Cook/Kreeger	16	calves	IM biobullet	1×10^{8}	Wyoming	vaccination challenge
Cook/Kreeger	16	calves	IM biobullet	1×10^{9}	Wyoming	vaccination challenge
Cook/Kreeger	10	bull calves	IM	1×10^{9}	Wyoming	safety
Thorne	70	pregnant	IM	1×10^{9}	Wyoming	pathogenesis
Thorne	7	pregnant	IM	1×10^{11}	Wyoming	pathogenesis
Kreeger/Cook	33	mixed*	oral	$1 \times 10^{9**}$	Wyoming	safety
Bison						
Olsen	6	calves	subcutaneous	1.2×10^{10}	Iowa	vaccination challenge
Olsen	10	calves	subcutaneous	1.2×10^{10}	Iowa	vaccination challenge
Olsen	20	calves	subcutaneous	1.5×10^{10}	Nebraska	safety

TABLE III-1 RB51 Vaccine Biosafety Tests (continued)						
Author	n	Animals	Route	Dose	Site	Study
Palmer	10	pregnant	subcutaneous	2×10^{10}	Montana	vaccination challenge
Rhyan/Olsen	12	adult	subcutaneous	1×10^{10}	Wisconsin	safety in bulls
Rhyan/Olsen	15	adult	subcutaneous	1×10^{10}	South Dakota	safety
Rhyan/Olsen	15	adult	biobullet	1×10^{10}	South Dakota	safety
Elzer/Davis	9	pregnant	subcutaneous	1×10^{9}	Texas	safety in pregnancy
Elzer/Davis	5	adult	subcutaneous	1×10^{9}	Texas	safety in females
Elzer/Davis	10	adult	subcutaneous	1.8×10^{10}	Texas	safety in bulls
Elzer/Davis	7	calves	subcutaneous	1.8×10^{10}	Texas	safety

*mix of adults, juveniles, and calves
**dose was given three times

Other Vaccines

The *B. suis* vaccine was developed in China. When given orally, vaccines prepared from *B. suis* strain 2 have been reported to reduce the incidence of porcine, bovine, ovine, and caprine brucellosis in China (Xie 1986); however, insufficient data are available to judge the usefulness of the vaccine. *B. neotomae*, a species from the desert woodrat, has been used to immunize caribou and reindeer against brucellosis. Although the extent of infection was reduced slightly, the number of abortions was not reduced (D. Davis, Texas A&M, pers. commun., 1997). Data on this vaccine in bison and elk have not been reported.

Mutants of *B. abortus* genetically engineered specifically for use in wildlife might prove effective in bison or elk. However, they have not been shown to offer substantial advantages in cattle (Cheville et al. 1993; Elzer et al. 1996). Several genes have been deleted from or added to *B. abortus*, but even though the resulting mutant survives in vivo, no clear advantage in vaccine efficacy has been established.

EFFICACY

Efficacy is the ability of an intervention to produce the desired beneficial effect. Live vaccines composed of organisms of diminished virulence have been most effective in reducing the incidence of brucellosis, tuberculosis, and diseases caused by most other facultative intracellular bacteria. In contrast, killed vaccines for this group of diseases have not proved efficacious. Several factors must be considered when assessing vaccine efficacy, including strain survival, route, dose, and age.

Strain Survival

When injected, a live vaccine must survive long enough to be immunogenic in the host. In vaccines composed of *B. abortus* for cattle, bacteria should survive in the lymph node draining the site of inoculation for at least 2 wk.

Four bison vaccinated with RB51 as young calves, allowed to mature, bred, and challenged with 10^7 CFU of *B. abortus* during the sixth month of pregnancy showed some degree of protection (Olsen et al. 1997): one aborted at 7 wk after challenge, and three completed a normal pregnancy with culture-negative calves. *B. abortus* was isolated from the parotid lymph node, bronchial lymph node, or uterus of two of the three female bison that had normal births.

Route

Vaccines can be delivered in several ways, including subcutaneously by hand injection, subcutaneously by dart, subcutaneously by biobullet, and orally.

Parenteral Injection

Subcutaneous vaccination is the preferred route of vaccination with B. abortus in most species. If it is done properly, and the vaccine is used before its expiration date, S19 and RB51 have been shown to produce an immune response in bison and elk. Conjunctival or intradermal vaccinations of bison and elk with S19 or RB51 have not been reported, but those routes have been used in cattle and were nearly as effective as subcutaneous vaccination, although not as practical.

Biobullet

The biobullet is composed of hydroxypropocellulose filled with a core of lyophilized, freeze-dried vaccine; it also contains stearic acid as a lubricant and calcium carbonate for weight. The 25-caliber bullet is fired from a compressed-air gun with an unrifled barrel. Although reported as inert, the hydroxypropocellulose bullet enters the tissue and produces trauma and some degree of foreign-body stimulation, and that might have some adjuvant effects on the vaccination process.

Oral

Oral vaccination has been tried to some extent in elk and bison. In a Wyoming study, RB51 was not efficacious in elk; there was no significant difference between vaccinates and controls in response to challenge. In contrast, a study of oral RB51 in elk in North Dakota showed some effect (Elzer et al. 1996). Problems with oral vaccination include control of dose per individual and inability to control the population to be vaccinated.

The experimental vaccination of cattle (heifers) with S19 has shown some protection: none of 20 orally vaccinated pregnant heifers challenged orally with strain 2308 at midgestation aborted, whereas 10 of 19 pregnant controls aborted (14 of the 19 were culture-positive) (Nicoletti 1981).

Dose

When any new vaccine is to be used, the dose required for effect and the margin of safety for bison and elk must be determined. Doses for commercial bison and elk follow the doses recommended for cattle. For greatest efficacy, the range of doses of vaccines should be established in bison and elk. High doses of both S19 and RB51 typically produce greater immune responses than low doses. Studies on RB51 in bison have not been extensive, but it appears that the most effective doses are close to those for cattle (Olsen et al. 1997).

Age

In general, young mammals will kill *B. abortus* and clear the organisms from their tissues more quickly than adults. The practical side of that is that in young calves *B. abortus* is less likely to be retained into puberty, when it infects the reproductive system. Data are needed in bison and elk differences in response to vaccination of calves, yearlings, adult females, and adult females with multiple inoculations. Data on cattle show little difference in host responses to vaccination with age from 3 to 10 mo (Manthei et al. 1950; Cheville et al. 1996).

Other Factors to Consider

Although it often is claimed that S19 and RB51 vaccines are less effective in bison and elk than they are in cattle, this has not been established clearly. Neither vaccine is 100% effective in cattle, and both have been shown to cause abortion when given in large doses, by inappropriate routes, or to pregnant animals (Palmer et al. 1996). Efficacy studies with similar numbers have not been done in bison and elk.

Wild, free-ranging animals differ from domestic animals in many factors that influence the effects of any vaccine. Among those factors are nutrition, shelter, veterinary care, and environmental stresses, such as food availability and the presence of geothermal energy. Other, more-subtle factors also could influence vaccine efficacy, such as the exaggerated metabolic changes that occur in bison and elk with handling, which can bring about capture myopathy.

The efficacy of live *B. abortus* vaccines varies with the age, sex, and genetic factors of the host. Although brucellosis in bison and elk closely resembles the disease in cattle, sheep, goats, and other ruminants, important species differences define the disease in a particular host. Variability in antibody and immune cell responses, in natural surface antigens, and in specific macrophage receptor molecules for *B. abortus* all are host-specific. For example, in cattle, normal serum has substantial antibrucellar properties, and it could be important to know whether that property exists in bison and elk.

Intercurrent infection at the time of vaccination can have a marked influence on the effect of a vaccine in individual animals. Infection of the host can enhance or diminish the effect of a vaccine, depending on the nature of the etiologic agent. Diseases that stimulate cell-mediated immunity might enhance the effect of some vaccines; for example, severe ringworm in cattle has been shown to increase markedly the antibody response to *B. abortus* (Cheville et al. 1992).

At the other extreme, systemic viral infections that replicate in the lymphoid system can suppress the effect of vaccines. In cattle, bovine viral diarrhea, bovine leukemia, and infectious bovine rhinotracheitis can have immunosuppressive effects. Similar immunosuppressive viral infections might occur in bison and elk and diminish the effects of vaccines.

Hormonal status, especially the activity of progesterone and other steroids, appears to affect how *B. abortus* is cleared from the host.

Duration of Immunity

The duration of immunity produced by vaccination of bison or elk is not known. Indirect evidence will come from surveys now under way in a Rhyan et al. study; data will be obtained on abortion and reproductive loss in vaccinated and nonvaccinated herds over time. It will be important to determine whether repeated vaccination is required.

Serologic Responses

RB51-vaccinated cattle that are subsequently infected with virulent *B. abortus* develop antibodies that react in standard serologic tests. Experimentally, RB51 vaccinates show increased titers in the standard tube test (STT), but, in contrast with responses in nonvaccinated controls, the titers drop progressively (Figure III-2).

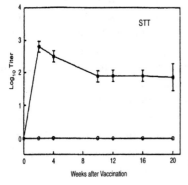

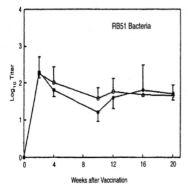

FIGURE III-2. Serologic responses of bison to vaccine RB51 (-o-, n = 6) or Strain 19 (-●-, n = 3) in the standard tube test (STT) or to RB51 dot-blot test (γirradiated RB51 antigen). Responses are presented as mean titer log10.

Vaccination of animals previously infected with virulent wild-type *B. abortus* could lead to unusual serologic responses. That possibility is important in animals vaccinated with RB51 and other vaccines that do not induce antibodies to lipopolysaccharide components of *B. abortus*. No data indicate that cattle infected naturally with virulent *B. abortus* develop an increased serologic titer in the STT after vaccination with RB51. Some data from field tests indicate that calves that have been vaccinated with S19 do not develop increased STT titers when given strain RB51 as adults. Bison probably react in the same manner.

Serologic responses of RB51-vaccinated bison remained negative at all times in all tests used. However, antibody responses to irradiated RB51 as an antigen were present in dot-blot tests 2 wk after vaccination. At 40 wk, two of six RB51 vaccinates had detectable antibody titers in the test for RB51, but S19 vaccinates did not.

BIOSAFETY OF VACCINES

No clinical disease or evidence of pathologic effect in tissue has been found even when large doses of S19 and RB51 vaccines were given to young bison calves subcutaneously. However, not all criteria of vaccine biosafety have been established or adequately evaluated (Table III-2).

TABLE III-2. Criteria for establishing biosafety in vaccines

- Clinical signs of acute disease do not appear after vaccination.
- Bacteria are not present in nasal secretions, saliva, or urine.
- Bacteria do not persist in the bloodstream for more than 3 days.
- Bacteria do not persist in lymph nodes for more than 16 wk.
- Evidence of humoral or cellular immunity is present 14 days after infection.
- No inflammation or chronic tissue injury appears.
- Neither placentitis nor abortion occurs in pregnant animals.
- Immunosuppression after 16 wk does not cause recrudescence.
- Bacteria recovered after 12 wk growth in the host are genetically identical with the vaccine strain.

Clinical Signs of Disease

A vaccine should not induce fever, loss of appetite (anorexia), or other clinical signs of disease. Neither bison nor elk vaccinated with S19 or RB51 develop significant clinical signs, and new vaccines also should not induce clinical manifestations. Lameness, which results from persistent infection of the joints (synovitis) with *B. abortus*, has not been reported. Vaccination-induced anaphylaxis, manifested by rapid development of shock and sudden death, occurs in a very small proportion of cattle vaccinated with S19; anaphylaxis has not been reported in wild mammals vaccinated with S19 or RB51.

Bacteria in Body Secretions

After vaccination, the vaccine strains of *B. abortus* should not appear in nasal exudates, tears, saliva, or other body secretions. Only sparse data are available, but S19 and RB51 have not been isolated from nasal swabs, saliva, tears, or urine of bison or elk. Milk has not been examined for excretion of S19, RB51, or other vaccine strains in bison and elk.

Bacteria in the Bloodstream

Live vaccine strains of *B. abortus* frequently can be isolated from the bloodstream in vaccinated cattle. It is common to be able to isolate the vaccine from large samples of blood for as long as 3 days after vaccination. In the blood, bacteria are typically associated with white cells. S19 and RB51 have been detected in bison in small numbers and only transiently in the bloodstream after vaccination. In one study of 10 bison given RB51, only one had culture-positive blood, and only one sample from that animal was positive (2 wk after vaccination) (Olsen et al. 1997).

Persistence in Regional Lymph Nodes

Live vaccines of *B. abortus* should not persist in lymph nodes draining the sites of vaccination for more than 16 wk. Replication or persistence beyond that time is associated with localization in the reproductive organs and mammary glands. In cattle, S19 and RB51 persist in lymph nodes after vaccination for times sufficient for development of immunity, but they are cleared before sexual maturity occurs. However, they might not be cleared from host tissues without risk of persistence to adulthood. Most bison vaccinated subcutaneously clear RB51 in 18 to 24 wk (Olsen et al. 1997). Biosafety experiments on RB51 in bison calves showed that replication of RB51 in lymph nodes draining subcutaneous sites of vaccination was greater than that in cattle. Three bison calves given 2.9 x 10^{10} CFU of S19 at the age of 3 months had S19 12 wk after vaccination but not 16 wk after vaccination (Table III-3). In bison at Ft. Niobrara National Wildlife Refuge, RB51 was present in all four lymph nodes tested 14 and 18 wk after vaccination, 22 wk after in one of four nodes, 26 wk after in three of four, and 30 wk after in none.

Immune Response

Vaccines given subcutaneously are rapidly taken up by the lymphatic system and induce a prompt antibody response. Antibrucella antibodies appear in the serum of bison about 2 wk after bison are given S19. Cutaneous delayed hypersensitivity after intradermal injection of brucellin or other evidence of cell-mediated immunity should also be detectable at about 2 wk after vaccination.

TABLE III-3 *Brucella abortus* in lymph nodes of bison after vaccination							
		CFU/g node tissue (wk after vaccination)					
Vaccine	n	1	2	4	6	10	16
RB51	6	667	12,769	7,337	165	174	9
		104	19,637	7,450	3,001	68	33
S19	3	104	131,967	7,746	1,216	324	0

Source: Olsen et al. 1997

Absence of Inflammation or Chronic Tissue Injury

No long-term tissue injury should be associated with vaccination; for example, chronic injury to the joints, brain, or other organs should be ruled out in biosafety tests. One consequence of infection with virulent *B. abortus* is a depletion of lymphocytes in the lymphoid tissues. This tissue destruction should not occur in *B. abortus* strains used for vaccines, and neither S19 nor RB51 has been shown to produce significant lymphoid destruction (Figure III-3).

Capacity to Induce Abortion

The capacity of a vaccine to be cleared by a calf before sexual maturity is critical with respect to causing abortion. When given to pregnant cattle and bison, all commercial live brucellosis vaccines that are designed for cattle have the capacity to infect the placenta and cause abortion. Placental lesions induced by vaccines cannot be differentiated pathologically from those induced by virulent field strains (Figure III-4). A dose of 5×10^8 CFU of S19 vaccine given to pregnant bison in the second trimester of pregnancy caused abortions in 58% (Davis et al. 1991); the same dose in cattle has been said to

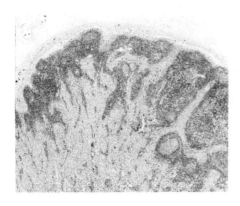

FIGURE III-3. Histology of a superficial cervical lymph node of a bison given Strain 19 vaccine 16 wk previously. No pathologic change. Germinal centers are present and medullary areas are large (Olsen et al. 1997).

induce fewer abortions (Nicoletti 1977). However, those two experiments were not critically compared. In the bison study, one S19-vaccinated cow aborted during her second pregnancy; that suggests that chronic vaccine infections can occur in bison.

Although RB51 can infect the bison placenta, it appears to be less abortigenic than S19 in bison and cattle, perhaps because of its diminished cell wall lipopolysaccharides. At high doses, RB51 vaccine has a tropism for the placenta and can cause abortion. Ten pregnant bison on a Montana ranch were vaccinated subcutaneously with 10^9 CFU of RB51. Two animals were sacrificed before 68 days; both showed vaccine-induced placentitis. Two abortions occurred—at 68 and 107 days after vaccination. The placenta of the aborting cows had placentitis associated with the presence of RB51 (Figure III-4).

Experimental Recrudescence

One danger in brucellosis research is the failure to detect live organisms when very few are present in tissues. That single *B. abortus* cells persist in a chronically infected animal in some "vegetative state" that precludes culture never has been proved. Use of new polymerase chain reaction technology to identify one organism in tissue might be useful. Another technique to show that live *B. abortus* cells are present is to treat an animal with dexamethasone or some other immunosuppressive treatment that will reactivate or recrudesce bacterial growth (when *B. abortus* cannot be isolated from chronically infected animals) and thus allow the organism to be detected. That has been used in cattle, but bison studies would be needed to develop a system of reactivation of bacterial growth.

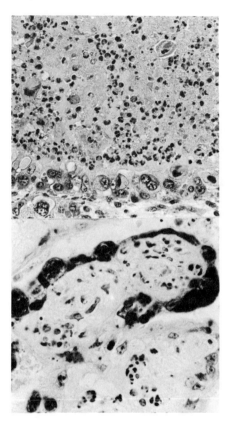

FIGURE III-4. Lesions caused by vaccine RB51 in the placenta of an aborting bison. A. Section of placentome 5 wk after vaccination with Strain RB51. Arcade zone and trophoblast layer contain large epithelial cells bearing *B. abortus*. B. Strong immunoreactivity for strain RB51 antigen in trophoblast epithelial cell cytoplasm (arrow); alkaline phosphatase (Palmer et al. 1996).

Genetic Stability

In establishing biosafety, the vaccine strain isolated after infection in the animal for several weeks must be genetically identical with the strain in the vaccine; for example, the bacterial strain must not mutate or revert during replication in the host. S19 has been shown to be stable in cattle after field use for more than 6 decades. RB51 has also been proved to be genetically stable in experiments in cattle, goats, and mice.

VACCINATION PROGRAM SUCCESS

The Greater Yellowstone Interagency Brucellosis Committee has agreed that "vaccination of bison and elk would achieve short term goals of brucellosis control and reduced risk of transmission to cattle." It is unlikely that a vaccination program for bison would succeed without a preliminary or concomitant program for elimination of brucellosis in elk. Re-emergence of brucellosis in a free bison herd is likely to occur if two conditions remain: brucellosis in winter elk feeding grounds is not controlled to prevent reinfection of bison from elk, and bison populations remain high.

Although it has been stated that brucellosis cannot be eradicated from free-ranging wildlife without eradicating the wildlife, it is certain that brucellosis can be eliminated from YNP with combinations of vaccination and culling. However, until a long-term controlled vaccination study is done, no assumptions about whether brucellosis can be eliminated by vaccination should be made. One measure of success would be a reduction in rates of abortion that would directly result in diminished bison-to-cattle and bison-to-bison transmission. Bison are more likely to prevent the long-term elimination of brucellosis from elk than vice versa.

IV
REDUCING THE RISK OF TRANSMISSION FROM WILDLIFE TO CATTLE

This chapter looks at approaches to reducing the risk of transmission from wildlife to cattle and reviews previous vaccination efforts in state and national parks and national wildlife refuges. Because any control or eradication effort will involve some degree of vaccination, this chapter reviews the difficulties involved in vaccine delivery and the effects of cattle vaccination on control efforts. Eradication efforts necessarily will include a test and slaughter component, and that component is examined for the effects on genetic diversity. And finally, this chapter looks at the prospects of natural regulation and successful brucellosis control.

PREVIOUS BISON-VACCINATION PROGRAMS IN NATIONAL AND STATE PARKS

In the 1960s, government regulations were devised to regulate interstate and intrastate movement of bison, and programs were developed to eradicate brucellosis in bison in some national parks and wildlife refuges. Bison herds managed with a herd plan were generally successful in eliminating brucellosis. Data on those cases are limited to memoranda from state and federal groups concerned with developing programs for brucellosis control and eradication (Gilsdorf 1997).

Wind Cave National Park and Custer State Park in South Dakota are adjacent to each other, and their bison herds had intermingled. High seropositive rates indicated both herds were infected with *B. abortus*. Brucellosis in adjacent cattle herds was being eradicated and had been eliminated by 1963 through vaccination, testing, and removal of reactor cattle. In Wind Cave, serologic testing for brucellosis in 1945 revealed 85% seropositivity; in 1960, a group of bison tested in Wind Cave had a reactor rate of 56%. Custer State Park had a 47 % reactor rate.

In April 1961, state and federal animal-health officials met with bison managers of both herds, and a herd plan was devised and agreed on. The plan included blood testing of the entire herd of adults and calves, immediate removal of reactors or permanent identification of reactors with later disposal, and continuing calfhood vaccination with S19. The Wind Cave and Custer bison herds were separated by a fence.

In Wind Cave, the program followed lines of "natural management," and facilities for active control of the herd of 250 bison were not built. In the first blood testing in 1964, 37% of the bison were seropositive (Table IV-1); by 1985, the herd was seronegative. The combination of vaccination, serologic testing, and management with removal of reactor bison allowed Wind Cave National Park to eliminate brucellosis in 21 years.

The program for the bison in Custer State Park followed lines of a commercial ranching operation. Capture facilities were built in 1960-1961. The first herd test, in the winter of 1961, found 119 reactors in 248 bison tested (Table IV-1). Bison were culled annually and sold or sent to abattoirs. All bison calves and yearlings were vaccinated annually. In 1967, the number of bison tested was increased to 2,110; the reactor rate was 5%. In 1973, the herd was seronegative, and in 1974 the park managers reduced the herd size from 1,750 to 1,000. Brucellosis had been eliminated in 10 yr, even though not all bison were tested each year.

The U.S. Fish and Wildlife Service's Wichita Mountains Wildlife Refuge located in Comanche County, Oklahoma, had vaccinated its bison and longhorn cattle for brucellosis since the 1940s. Bison tested in October 1963 were seronegative, but in 1964 brucellosis suspects in elk and bison were found in the fall roundup. The origin of the disease is not known. The program developed for this herd included the following steps: blood-test all bison over 1 yr old in the fall of 1972, slaughter all bison that could not be gathered, send all test-positive bison to slaughter and collect tissues for isolation of *B. abortus*, divide bison herd into isolated groups on different pastures, conduct a complete herd test in the fall of 1973, discontinue vaccination of bison calves in 1973, and test another species for brucellosis. The Refuge also reduced the size of the herd from 781 to 345 in 1973. It took 8 yr to eliminate the disease, and the herd was considered free of brucellosis in May 1974.

Table IV-1. Bison seropositivity rates in parks that eliminated brucellosis. (Source: Gilsdorf 1997)

Year	Reactors/No. Tested	
	Custer State Park	Wind Cave National Park
1961	119/248 (48%)	
1962	20/141 (14%)	
1963	?	
1964	2/84 (2%)	81/220 (37%)
1965	0/16 (0%)	41/175 (23%)
1966	20/905 (2.2%)	16/173 (9.2%)
1967	113/2,110 (5%)	12/185 (6.5%)
1968	53/2,493 (2%)	7/194 (3.6%)
1969	3/1,335 (0.2%)	7/282 (2.5%)
1970	7/1,439 (0.5%)	1/75 (1.3%)
1971	1/1,142 (0.09%)	1/146 (0.7%)
1972	12/1,379 (0.9%)	1/146 (0/7%)
1973	0/108 (0%)	?
1974	?	2/120 (1.7%)
1975	0/172 (0%)	
1977	0/237 (0%)	
1979		12/185 (6.5%)
1982		3/128 (2.3%)
1983		15/264 (5.7%)
1984		7/337 (2.1%)
1985		0/225 (0%)
1986		0/217 (0%)
1987		0/205 (0%)
1989		0/191 (0%)

APPROACHES TO CONTROLLING
OR ELIMINATING BRUCELLOSIS IN YNP

Numerous approaches to controlling or eliminating brucellosis from the Greater Yellowstone Area (GYA) have been identified. Some are theoretical, some are experimental, and others are technically possible. All should be considered for short- and long-term solutions.

• *Vaccination alone* with a vaccine that is protective will reduce but not eliminate *B. abortus* from the GYA (Peterson et al. 1991a). Development and use of an efficacious vaccine could greatly reduce the prevalence of brucellosis in the GYA.

To be successful in bison, vaccination must be accompanied by prevention of contact with infected elk, and reduction of brucellosis in elk by reducing feeding-ground concentrations. Artificially controlling population growth in bison would make administering programs to eliminate brucellosis easier. The disparity of seroprevalence between feeding-ground elk and the northern herd suggests that exposure to infected material on the feeding ground is the driving force maintaining infection in elk. Management strategies to disperse elk from the feeding grounds for the 3 months before calving combined with an intensive vaccination program might eliminate the disease from elk. Discontinuing winter feeding of elk would eliminate the problem of elk congregating but have the consequence of drastically reducing the number of elk in the Jackson area.

Vaccinating a high-enough proportion of elk is problematic because they are widely distributed, and foci of infection are numerous in Wyoming and the National Elk Refuge (NER). Vaccination and gradual removal of feeding grounds as elk foci would probably allow the gradual natural extinction of the disease in elk (K. Aune, Mont. Dept. Fish, Wildlife and Parks, pers. commun., 1997).

• *Vaccinating cattle and bison* would make the risk of transmission from bison extremely low under current conditions.

• *Spatial and temporal separation of cattle and bison* would be a good first step toward risk reduction. Regional surveillance and monitoring of surrounding cattle in Montana, Wyoming, and Idaho might be required for early phases of any program.

• *Vaccination combined with herd management*—including culling and test-and-slaughter procedures, close surveillance and monitoring of disease prevalence, and spatial and temporal management of wildlife—could be used to eliminate the disease in bison. Vaccination alone would have to be continued indefinitely, but if it were combined with a test-and-slaughter program, brucellosis potentially could be eradicated in the GYA over time. The time required for eradication would be contingent on vaccine effectiveness, the slaughter rate, and efforts to reduce the population. Case studies of Custer State Park have shown that eradication would take 10-20 yr (Gilsdorf 1997).

"In theory and in practice, vaccination combined with test and slaughter is effective, second only to depopulation, in eradicating brucellosis from cattle" (T. Kreeger, WGFD, pers. commun., 1997).

One program suggested for control and eventual eradication is vaccination and test and slaughter of bison and elk with restriction of perimeter cattle herds to steers and monitoring of peripheral cow herds. Key elements of the program are listed in below in order of importance; the first three were considered essential (S. Amosson, Texas A&M, pers. commun., 1997):

- Test and slaughter of all segments.
- Perimeter control through use of steers (or full vaccination of cow herds).
- Vaccination of bison herds.
- Vaccination of elk herds.

Opinions differ as to the likelihood of successful outcomes of the various programs. The likelihood of reduction of abortions in bison and elk and reduction in transmission to cattle seems high. The likelihood of eradication of *B. abortus* in the short term is low but would increase with appropriate levels of funding and an adequate vaccination program (E. Williams, Univ. Wyom., pers. commun., 1997).

R. Mead (State vet., Wash., pers. commun., 1997) compared different approaches:

Pros and Cons of Approaches to Control	
Vaccination only	Vaccination with test-and-cull program
slow	faster
expensive	more traumatic to the animal
more acceptable to the public	less acceptable to the public
requires more research	greater risk of human injury
adverse impact on GYA management	

Cost is a major problem. For the past decade, the annual federal budget for the U.S. Department of Agriculture program for brucellosis eradication in cattle has been about $65 million. Added to that are the indirect costs of state programs and producer vaccination efforts, and production losses must be considered. Brucellosis eradication in cattle is a costly process.

In combination with vaccination, alternative methods might be acceptable and effective. Testing and neutering of seropositive animals is an option that has not been considered. Neutered animals are not likely to spread brucellosis. However, surgical neutering might not be publically acceptable and might have undesirable effects on herd behavior. Immunocontraception to suppress persistence of *B. abortus* in bison should be examined, but current technical barriers, especially with delivery systems, and the potential for introducing genetic selection makes it unacceptable at present. Pregnancy reduction has been achieved in captive deer (Turner et al. 1996a) and free-roaming feral equids (Turner et al. 1996b), and new techniques also provide promise (Miller et al. 1997). Whatever methods are used, results will depend on the intertwined effects of bison and elk populations, predator numbers, food supply, and weather.

FIELD DELIVERY OF A VACCINATION PROGRAM FOR YNP BISON

Given the high seropositivity rate in YNP bison (about 50%), several people have noted that a test-and-slaughter program to eliminate brucellosis would differ little from a depopulation program. Neither depopulation nor a test-and-slaughter program alone is likely to be publicly acceptable in YNP. More realistic is the implementation first, of vaccination to reduce the seropositivity rate to a low level, and then, when the numbers that have to be removed are small, a test-and-slaughter program. That strategy could be conducted within the framework of an adaptive management approach.

A program of vaccination for bison in the field in YNP in all likelihood will have to be conducted under several constraints. Most bison populations have been managed for many years by rounding up in specially designed corrals, and any incorrigible individuals that could not be managed were shot. Administering a brucellosis-elimination program similar to that used for domestic livestock (vaccination in conjunction with test and slaughter) is feasible in those cases. But rounding up has the consequence of some artificial selection for domestication because wildness and intractability, salient traits of wild bison, are disfavored. Those are important traits to retain in YNP bison, one of the few herds where it is feasible to maintain natural behavior, so rounding up is not likely to be acceptable. In addition, the construction of facilities necessary to handle bison would detract from the natural aura of the park and might have detrimental effects on the park ecosystem.

Consequently, whatever vaccine is developed probably will have to be applied to free-roaming bison. As noted above, brucellosis was eliminated in one herd (Wind Cave National Park) with vaccination and test and slaughter. Most research veterinarians think that the vaccine should be developed first and a method of application found later. However, in view of the constraints likely to pertain to application in YNP, it seems prudent to keep the field-delivery problems of YNP in mind, for they might influence the characteristics required of the vaccine.

Vaccine Delivery in Food or via Injection

Vaccine is likely to be delivered as bait, in food, or via remote injection. All have serious problems in the context of YNP bison. Putting vaccine in bait or artificial food has the drawback of not allowing control of doses. Dominant animals are likely to get multiple doses; other animals might get none. Bulls would be treated with cows. Perhaps low doses given ad lib would result in all animals receiving a common dose per body mass, but that seems optimistic. An alternative strategy would be to give the vaccine over a short time with feed spread to allow consumption by all individuals. Nevertheless, control of dose by feeding strategy would be difficult. Furthermore, it would be difficult to prevent other species from eating treated feed. Imaginative approaches using genetic engineering to put the vaccine in native plants have been proposed (D. Sands, Mont. State Univ., pers. commun., 1997), but that technology is probably many years away and will be subject to the same problems of control of dose as the artificial feed route or worse. In addition, serious issues of ecologic and evolutionary consequences probably are best not worked out in YNP, the crown jewel of the U.S. national park system.

Remote injection is probably a more realistic approach. However, bison are not fed in the winter and are not as approachable as elk that are currently vaccinated on winter feeding grounds with biobullets. The range of biobullets is very short, and this is not a good technique for wary animals. Syringe darts have a greater range, and they can be fired from the ground or by helicopter. Disturbance is an issue. A vaccination program probably would best be conducted in winter when visitors are fewer, and perhaps it could be carried out away from the roads to which snow machines are confined. Whether unrecovered syringes are an environmental hazard would have to be addressed.

A major problem will be to distinguish treated animals from untreated animals. Temporary marking as is done with elk will be possible with biobullets or syringes; even then, the dark coat of the bison will make such marks more difficult to see. In a milling herd, the identification of treated individuals will pose a problem.

Directing the program at recognizable age or sex classes would reduce the number of animals that need to be vaccinated each year. Because of they are small and easy to recognize, calves are a favorable target group, except that it can be difficult to distinguish male from female calves (females are the likely sex to be vaccinated) in the field. Yearling cows might have greater potential in being present in small numbers, distinguishable from bulls, and recognizable by experienced field personnel.

Venereal Immunization

The vaccination of bulls to immunize cows has been suggested by some. Venereal immunization of cows by purposeful infection of dominant breeding bulls at the right time so that the infection is at its peak in the breeding season theoretically would immunize most of the adult females in the population by vaccinating only a small subset of the population. Old dominant bulls are most recognizable by size and often recognizable individual marks. According to the study of Berger and Cunningham (1994; J. Berger, U. Nev., pers. commun., 1997), 6-yr-old and older bulls do most of the breeding, and they constitute about only 6-7% of the population. The low number of vaccinations would be more feasible to deliver in the field and would intrude on the segment of the population that would manifest the least demographic consequences of disturbance.

There is no evidence to suggest that venereal immunization would be effective. Vaginal epithelium is a strong barrier and lacks the macrophages that make the uterus susceptible to infection. Failure of venereal transmission of *B. abortus* in cattle and other species is based on that difference and underlies the failure of experimental intravaginal inoculation to transmit *Brucella* spp. The number of males that become infected, the percent of infected males that excrete *B. abortus* in semen, and the failure of bulls with testicular lesions and pain to breed all make venereal transmission unlikely. In addition, new vaccine strains of *B. abortus* have the attribute of reduction of reproductive tract tropism. Selecting the vaccine strain of *B. abortus* for appropriate characteristics (such as higher rates of shedding of *B. abortus* in the semen), timing of injection in the bulls, or similar refinements of the technique might overcome these problems.

VACCINATION OF CATTLE

Given the difficulties of vaccinating bison, the most workable method of reducing the risk of transmission of brucellosis from bison and elk to cattle in the GYA is vaccination of cattle. Cattle are already rounded up and handled, so the major impediment to uniform vaccination against brucellosis is the associated cost. Most cattle in the region already are being vaccinated for brucellosis, and this program is the most cost-effective way of reducing potential transmission from wildlife in the short term. Vaccination is required in Idaho and strongly recommended in Montana and Wyoming.

Until a program of elimination is in the implementation stage, cattle vaccination should be universal in the area surrounding the GYA.

LIMITING CATTLE NEAR PARK BORDERS TO STEERS

The presence of geographic barriers that reduce the spread of brucellosis by limiting contact of infected bison and elk in the GYA with susceptible cattle clearly is important. One approach based on this principle is to reduce contact by making the first line of contact a population of cattle that has a reduced likelihood of maintaining *B. abortus* in the herd. Limiting cattle near YNP borders to steers or spayed heifers could lower the risk of transmission in the treated animals.

Castrated males and spayed heifers are unlikely to transmit brucellosis. Removal of the testes and ovaries deletes the source of gonadal hormones that initiate reproductive growth at puberty and maintain the reproductive system in adulthood. Although the animals might become infected, they will not transmit the disease when living and do not develop tissue titers of bacteria that can sustain the disease in nature. Certainly steers and spayed heifers do not transmit brucellosis through abortion or its byproducts, but the requirement for sexual maturity and the presence of gonads in transmission has not been clearly established.

Opposition has been voiced to limiting cattle production around the GYA to steers. Limiting cattle production will not eliminate brucellosis; elk will remain throughout a large landscape. Some also believe that this method wrongly places responsibility on agricultural segments of society.

A related plan that has merit and could be carried out immediately is to establish perimeter zones in which animal populations are monitored in progressively vigorous ways. Zones established nearest the GYA would have increased disease surveillance, vigilant monitoring, vaccination, and contact-reporting programs. Implementation of such a perimeter-zone strategy should include collection of serologic data in cattle vaccinated with RB51. This would more clearly establish whether transmission of *B. abortus* actually is occurring in the GYA.

EFFECTS OF TEST-AND-SLAUGHTER PROGRAMS ON GENETIC DIVERSITY

Reduction of wild populations to low numbers for any reason raises concerns over loss of genetic diversity (Denniston 1977; Frankel and Soulé 1981). Bison in YNP contain lineages that go back without interruption to the aboriginal stocks in the area (Meagher 1973). In fact, the YNP population is the only extant bison population that has not been derived solely from stocks held in captivity at some point in their history. Plains bison *(Bison bison bison)* stock was introduced to YNP (Meagher 1973), but if any current population is likely to contain unique alleles from the original bison *(Bison bison athabaska*, which occupied the valleys in the Rocky Mountains), it is the YNP herd. Consideration of minimal numbers must include genetically effective population size, which is influenced by sex ratio, breeding behavior, the number of nonbreeding individuals, and other factors. It can be substantially less than actual population size in a polygynous species like bison. Berger and Cunningham (1994) calculated effective population size in bison to be 21-46% of actual size, depending on the variables included in the formula applied. For example, if the goal were to maintain an effective population of at least 500 bison for gene conservation, an actual population of 1,087-2,381 would be required.

Protein-electrophoresis data suggest that the YNP herd and the Wind Cave National Park herd have the highest heterozygosity (a measure of genetic diversity) among the 12 public herds of bison in the United States (Stormont 1993). However, DNA studies of the YNP bison, using both mitochondrial DNA (inherited only through the female lineage) and nuclear DNA (microsatellites—a sensitive measure of genetic changes over time), revealed no unique alleles in that population (J. Derr, Texas A&M, pers. commun., 1997). The lack of unique alleles in YNP might indicate that mountain bison were not very different from plains bison or that genetic diversity was lost because of the bottlenecks and long periods at small population sizes characteristic of the population (see Figure II-2). Considering the influence of effective population size, the number of bison was small in the early years. Alternatively, it might be that because bison stocks have been mixed frequently, including the movement of bison from YNP to other populations, their genes are represented in other herds. Whatever the explanation for the apparent absence of unique alleles, the DNA evidence suggests that conservation of genetic diversity is not a major issue in the management of the YNP bison. Technical decisions can be based on mainly demographic criteria. However, only a small part of the genome has been analyzed, and prudence dictates that minimal effective size be considered in any program of brucellosis eradication.

NATURAL REGULATION AND BRUCELLOSIS CONTROL

The analysis of movements of bison and elk outside YNP highlights the importance of the park's policy of "natural regulation" of ungulates in relation to the possibility of transmission of *B. abortus* to domestic livestock. Given that bison and elk populations are large, they will continue to move out of the park; that is especially true of bison in years with hard winters. It cannot be determined with precision what the transmission risk is, because with current knowledge, it is too small to measure with accuracy. However, whatever the risk, it will be increased by more frequent movement of greater numbers of bison and elk beyond park boundaries. Whether the increase in risk is trivial or important depends on how the epidemiologic evidence is interpreted. In any event, the YNP policy of natural regulation influences the probability of transmission of *B. abortus* from wildlife to cattle and therefore must be considered in this study, although specific recommendations regarding the policy are beyond the study charge.

Natural-regulation policy, particularly as it pertains to the northern YNP elk herd, has been controversial (Houston 1982; Chase 1986; Kay 1990). As with brucellosis, the science is insufficient to settle arguments over whether it is wise. Critical tests are difficult because the issues are linked with larger patterns of nature that are not readily reduced to a research hypothesis. The "experiment" is conducted by nature, lacks controls and replications, and yields only one set of data points per year. Each side in the debate interprets the results with reference to its own position.

Natural regulation is a useful label for the controversy, but the real issue is human intervention: should it occur? If so, to what degree and when? The mandate of the National Park Service in large wild parks is to maintain wildlife in as natural a state as possible. If that mandate cannot be carried out in a park as large as Yellowstone, it has little prospect in any of the other national parks in the lower 48 states. Given the ubiquitous alteration of landscapes by human pressures outside the parks, YNP and a few other large parks are the main remaining baseline areas where the course of nature can be observed (Sinclair 1983; Arcese and Sinclair 1997). They are the controls for the national and global human experiment. No one would seriously argue that YNP—with its infrastructure of roads, accommodations, and millions of visitors—is in a natural state. It is not immune to edge effects along its borders or to regional or global phenomena. It is not a complete ecologic entity, as indicated by larger designations, such as the "Greater Yellowstone Area" and the "Greater Yellowstone Ecosystem." Nevertheless, retaining the ecologic integrity of the park requires setting limits and not intruding where nature can manage without human intervention. In fact, many historical human intrusions, such as husbandry of ungulates and attempted elimination of large predators, are lamented by most conservation scientists and require restoration efforts, such as the recent reintroduction of the wolf. Those interventions were deemed advisable at the time, and they should stand as stark reminders of the limits of knowledge and understanding—then and now. Decisions to intervene should be supported by clear and compelling evidence and a consensus of experts that they are necessary.

On the other side of the issue is a long history and practice of managing ungulate populations to meet prescribed goals. Virtually every bison and elk population in the country outside YNP is managed to some degree, and they were managed inside YNP at an earlier time—bison until 1967 (Meagher 1973) and elk until 1969 (Houston 1982). Neither has been managed in YNP since the implementation of the natural-regulation policy in 1969. Elk that migrate out of YNP in the fall and winter are hunted in the surrounding states, and elk are still managed in the Grand Teton National Park area (Boyce 1989; Toman et al. 1997). Management goals for ungulates usually are to stabilize the population within some range deemed to be in the best interest of the health of the population and to maintain some state of vegetation that is judged to be desirable. To some critics of YNP policies, the need to control ungulates to prevent irruptive population behavior and its consequent detriment to vegetation is a guiding principle in ecology. Obviously, the knowledge and technical capability are available to manage bison and elk to stabilize their numbers inside YNP at some upper limit. The important question, therefore, is not whether we can, but whether we should do so.

The debate over bison and elk management in YNP has taken place before a larger backdrop of a shift in paradigms in the field of ecology. Stated simply, the earlier paradigm was based on equilibrium states—the "balance of nature"—whereas recent emphasis has been on nonequilibrium fluctuations over time and space (Simberloff 1982; Pickett et al. 1992). According to the latter paradigm, fluctuations are the norm, and any apparent balance is an artifact of averaging over space or long periods. The shift was prompted by the growing body of evidence that variation, rather than similarity, across landscapes is the predominant characteristic of nature and that analyses of longer time series were showing the same to be true over time (for example, Sousa 1984; Pickett and White 1985; Whitlock 1993; Reice 1994; Russell 1994). At the same time, little hard evidence could be marshaled to support the equilibrium view.

Traditionally, irruptive population behavior in ungulates (exceeding average carrying capacity and suffering declines due to starvation) was viewed as exceptional and attributable to the actions of humans: confinement by fencing or development, introductions to new areas, removal of predators, and so on. Such irruptions showed strong associations in time with human effects on ecosystems, and certainly the actions of humans contributed to the degree of irruption, even if not they were not the cause. But close examination of the evidence supporting those cases shows that it is often deficient (Caughley 1987; McCullough 1997). In fact, there is little solid evidence to support the traditional view. The population estimates in most cases were largely guesses, and contributing variables were not measured. Even if the traditional view is accepted on logical grounds, it cannot be assumed that because human intervention can produce irruptions, nature does not. Actually, the earlier irruptions can be characterized better as comparisons between unhunted populations and hunting-controlled populations than as human intrusion versus the natural order. Human intervention was ubiquitous; it was, and still is, hard to find places where natural behavior could be observed as a "control." That ungulate populations in undisturbed nature tended toward an equilibrium based on interactions with predation and resources was inferred from the assumption that natural predators caused effects equivalent to human hunting in the cases in which hunting had stabilized populations (Leopold 1933, 1940). Early results in the moose and wolf populations in Isle Royale National Park gave initial credence to the equilibrium view (Mech 1966, 1970; Allen 1979), but later results dispelled any semblance of equilibrium (Peterson and Page 1988; McLaren and Peterson 1994).

Despite restoration efforts in parks and reserves, finding natural areas from which to obtain time-series data to address the behavior of ungulate-predator-resource-climate interactions continues to be difficult. Control and replicate areas on a scale necessary for ungulates and their predators are not available, and treatments are not possible if natural management is pursued. Consequently, long series of years are necessary to observe the effects of "natural experiments" that occur by chance. Interpretations are based on correlations, and, because nature seldom conducts clean experiments, cause and effect often are complicated by covariance of variables. The difficulties are similar to those of determining whether global warming is occurring and, if so, whether it is due to natural processes or anthropogenic carbon dioxide emissions. The answers will come, but they are not in the immediate offing. Despite the difficulties, a growing body evidence from largely intact natural areas seems to indicate that ungulate populations commonly fluctuate over considerable numbers.

Abundant evidence from YNP indicates that neither bison nor elk conform to a stable equilibrium model. Both species have shown a persistent tendency to increase to the limits of the environment. That the northern elk herd shows a dynamic equilibrium should not obscure the fact that the amplitude of its population fluctuation over time is considerable. The elk herd shows a dynamic equilibrium in that the rate of increase has declined in a density-dependent manner at high population numbers (Coughenour and Singer 1996; M. Taper, Mont. State Univ., and P. Gogan, USGS, pers. commun., 1997). The dynamic equilibrium mean carrying capacity (that is, where elk population growth rate equals 0) appears to be about 14,000-18,000 elk (Houston 1982; Merrill and Boyce 1991; M. Taper, Mont. State Univ., and P. Gogan, USGS, pers. commun., 1997). Bison, in contrast, have not yet shown evidence of natural regulation over the range of numbers recorded, and their geographic expansion has already exceeded the boundaries of YNP. Natural regulation of bison in YNP appears to be unlikely. Control of bison numbers presents difficult choices that had to be addressed in the recent past and probably will have to be addressed again, independently of the brucellosis issue. Although brucellosis has catalyzed the recent controversy, the fundamental issue is the need to respond to burgeoning bison numbers that are overflowing park boundaries.

Growth of bison and elk populations has been expressed in the presence of native predators other than the wolf. Whether the recently reintroduced wolves, whose population has grown quickly to about 100, will have an appreciable effect on bison and elk population growth remains to be seen. A computer model using a wolf population of 76 predicted an elk population reduction over 100 yr of 15-25% but no evidence of imposing an equilibrium (Boyce 1993). The same model predicted a bison population reduction of less than 10%. But judging by the experiences with moose on Isle Royale (Peterson and Page 1988; McLaren and Peterson 1994) and other ungulates elsewhere (for example, Ballard et al. 1987; Gasaway et al. 1983, 1992; NRC 1997), it is questionable whether wolves will impose a stable equilibrium on either bison or elk in YNP. No stable equilibrium is apparent on Isle Royale, with almost 40 yr of data probably the best documented case of interaction between populations of a natural, large ungulate and wolves. Mech et al. (1987) noted that winter snow is the most important variable controlling the number of moose on Isle Royale and that wolves had a relatively minor effect.

A concept in ecology that well could be applied to the YNP elk and bison population issue is that of source-and-sink dynamics (Pulliam 1988). It is based on the common observation that some habitat patches favorable to a species result in production of new individuals greater than can be supported in the habitat patch; subsequent population pressure results in the dispersal of part of the population into surrounding poor habitats where reproduction and survival are low. Thus, a source population is a consistent exporter of individuals, whereas a sink population cannot maintain itself except for the continuous influx of individuals from the source. The net outcome for both areas is a dynamic equilibrium, with the majority of reproduction happening in source habitats and the majority of mortality in sink habitats. The concept would appear to apply well to the YNP elk and bison situation. As long as the natural-regulation policy is followed, increasing elk and bison populations will stretch the winter capacity of YNP, and, at least in harder winters, animals will be forced out of the park. The incompatibility of bison with developed areas and private lands will require either culling or relocation; both have the demographic consequence of removing the animals from the system. Because it is a source habitat, YNP can continue to be managed according to the natural-regulation policy. The sink habitat outside the park can be the area where adjustment is applied by the combination of relocation and mortality that is compatible with brucellosis containment and public acceptance. (The sink habitat in the GYA is a sink not because of inadequate habitat quality. It is a sink because of high mortality, in this case, mortality induced by humans.) The populations of bison and elk can be stabilized over the combined areas (the ecosystem) in a manner that duplicates or mimics a common system in nature without violating the mandates under which these lands are managed. In many ways, source-and-sink dynamics already applies to the management of YNP elk and bison.

At minimum, source-and-sink dynamics has heuristic value. For example, it puts various proposals about management of federal lands (mainly by the Forest Service) into a different perspective. It has often been suggested that bison and elk should be favored over other uses, particularly livestock grazing, on these lands. That proposal is attractive for its inherent appeal of contributing to the conservation of bison and elk, and the lands are already in public ownership. The perspective of source-and-sink dynamics, however, reveals two drawbacks of this approach.

First, it assumes that the additional area will contribute to natural regulation. That might be true for northern-range elk, whose population shows evidence of regulation. However, Coughenour and Singer (1996) and M. Taper (Mont. State Univ.) and P. Gogan (USGS, pers. commun., 1997) have noted that elk populations in the northern range previously have expanded in response to habitat increase. It is possible that they might do so again, although Coughenour and Singer (1996) note that additional suitable habitat might not be available. Bison, however, have shown no evidence of regulation, but only range expansion. The likely consequence of shifting the boundary of protection from YNP to surrounding public lands is that bison, and perhaps elk, populations will simply increase further, shifting the boundary to a new point—private lands—where even greater numbers of bison will have to be dealt with. Those limits need to be confronted unless our nation is ready to make a substantial commitment to acquire private lands for bison conservation. If such a commitment is to be made, it needs to be determined whether it should be made in the GYA where bison conservation is already near the potential of the ecosystem. Bison conservation might be better served if, for example, the commitment were directed to the Great Plains, the heartland of the aboriginal bison range.

Second, the sharp juxtaposition of source-and-sink areas maximizes the conflicts because the policies of one jurisdiction or the other will have to be compromised to some extent by the lack of a transition area. Establishment of buffer zones between parks or reserves and the surrounding lands used for agriculture or other purposes is a well-accepted approach in land planning (for example, Harris 1984; Western and Wright 1994). The buffer zone is an area in which management can facilitate the transition between goals of two contrasting land uses. In a source-and-sink model, then, the shift from favorable to unfavorable habitat (because of conflict with human land uses) is accommodated along a gradient in the buffer zone between protected and unprotected areas. Federal lands outside YNP could, and to some degree already do, serve that function. The buffer zones also could be linked with perimeter zones for brucellosis control discussed earlier.

ADAPTIVE MANAGEMENT

Because neither sufficient information nor technical capability is available to implement a brucellosis-eradication program in the GYA at present, eradication as a goal is more a statement of principle than a workable program. The best that will be possible in the near future will be reduction of the risk of transmission of *B. abortus* from wildlife to cattle.

Biologically sound wildlife policy can be developed most efficiently using adaptive management (Walters 1986; Lancia 1996, NRC 1997). An adaptive management approach that had research designed to provide data to reduce areas of current uncertainty should eventually give a more realistic assessment of the feasibility of eradication of *B. abortus* in the GYA. Adaptive management means conducting management activities as hypothesis tests, the outcome of which will direct the subsequent efforts to achieve the ultimate goal. Adaptive management is not just modifying management in light of experience; it is designing management intervention to maximize what can be learned from the experiments (NRC 1997).

REFERENCES

Allen, D.L. 1979. Wolves of Minong: Their Vital Role in a Wild Community. Boston: Houghton Mifflin. 499 pp.

Altmann, M. 1952. Social behavior of elk, *Cervus canadensis* Nelsoni, in the Jackson Hole area of Wyoming. Behaviour 4:116-143.

Alton, G.G., L.M. Jones, R.D. Angus, and J.M. Verger. 1988. Techniques for the Brucellosis Laboratory. Institute National de la Recherche Agronomique. Paris. 190 pp.

Anderson, C.C. 1958. The elk of Jackson Hole, a review of the Jackson Hole elk studies. Wyoming Game and Fish Bulletin 10. 184 pp.

Anderson, T.D., N.F. Cheville. 1986. Ultrastructural morphometric analysis of *Brucella abortus*-infected trophoblasts in experimental placentitis. Bacterial replication occurs in rough endoplasmic reticulum. Am. J. Pathol. 124(2):226-237.

Angus, R.D., G.M. Brown, and C.S. Gue, Jr. 1971. Avian brucellosis: a case report of natural transmission from cattle. Am. J. Vet. Res. 32(10):1609-1612.

Arcese, P., and A.R.E. Sinclair. 1997. The role of protected areas as ecological baselines. J. Wildl. Manage. 61:587-602.

Arman, P. 1974. A note on parturition and maternal behaviour in captive red deer (*Cervus elaphus* L.) J. Reprod. Fertil. 37(1):87-90.

Aune, K., and P. Schladweiler. 1992. Wildlife laboratory annual report. Montana Dept. Fish Wildlife Parks. Bozeman MT. 55 pp.

Aune, K., T. Roffe, J. Mack, and A. Whitelaw. 1997. Preliminary results on home range, movements, reproduction, and behavior of female bison in the Lamar Valley, Yellowstone National Park. (Abstract) Proc. Intl. Symp. on Bison Ecology and Management in North America. June 4-7, 1997. Bozeman, MT.

Ballard, W.B., J.S. Whitman, and C.L. Gardner. 1987. Ecology of an exploited wolf population in south-central Alaska. Wildl. Monogr. 98. 54 pp.

Bangs, E.E., and S.H. Fritts. 1996. Reintroducing the gray wolf to central Idaho and Yellowstone National Park. Wildl. Soc. Bull. 24:402-413.

Barmore, W.J. 1968. Bison and Brucellosis in Yellowstone National Park: A Problem Analysis. Yellowstone, WY.: Yellowstone National Park Research Library. 73 pp.

Bauer, E.A. 1995. Elk: Behavior, Ecology, Conservation. Stillwater, MN.: Voyageur Press, 160 pp.

Bercovich, Z., J. Haagsma, E.A. ter Laak. 1990. Use of delayed-type hypersensitivity test to diagnose brucellosis in calves born to infected dams. Vet. Q. 12(4):231-237.

Berger, J., and C. Cunningham. 1994. Bison: Mating and Conservation in Small Populations. New York: Columbia University Press, 330 pp.

Bork, A.M., C.M. Strobeck, F.C. Yeh, R.J. Hudson, and R.K. Salmon. 1991. Genetic relationship of wood and plains bison based on restriction fragment length polymorphism. Can. J. Zool. 69:43-48.

Boyce, M.S. 1989. The Jackson Elk Herd: Intensive Wildlife Management in North America. Cambridge; New York: Cambridge University Press, 306 pp.

Boyce, M.S. 1993. Predicting the consequences of wolf recovery to ungulates in Yellowstone National Park. Pp. 234-269 in Ecological Issues on Reintroducing Wolves into Yellowstone National Park. R.S. Cook, ed. Scientific Monograph NPS/NRYELL/NRSM-93/22. USDI National Park Service, Denver, CO.

Boyce, M.S., and E.H. Merrill. 1996. Predicting effects of 1988 fires on ungulates in Yellowstone National Park. Pp. 361-366 in Effects of Grazing by Wild Ungulates in Yellowstone National Park. F.J. Singer, ed. Technical Report NPS/NRYELL/NRTR/96-01, USDI National Park Service, Natural Resource Program Center, Natural Resource Information Division, Denver, CO.

Bricker, B.J., and S.M. Halling. 1995. Enhancement of the *Brucella* AMOS PCR assay for differentiation of *Brucella abortus* vaccine strains S19 and RB51. J. Clin. Microbiol. 33(6):1640-1642.

Broughton, E. 1987. Diseases affecting bison. Pp. 34-38 in Bison Ecology in Relation to Agricultural Development in the Slave River Lowlands, N.W.T. H.W. Reynolds, and A.W.L. Hawley, eds. Ottawa: Canadian Wildlife Service.

Brown, W.M., M. George, Jr., and A.C. Wilson. 1979. Rapid evolution of animal mitochondrial DNA. Proc. Natl. Acad. Sci., USA. 76(4):1967-1971.

Buck, J.M., G.T. Creech, and H.H. Ladson. 1919. *Bacterium abortus* infection of bulls (preliminary report). J. Agric. Res. 17(5):239-252.

Cariman, R.L. 1994. Wildlife-private property damage law. Land and Water Law Review. Vol XXXIX(1), p. 89-115. University of Wyoming College of Law.

Caughley, G. 1987. Ecological relationships. Pp. 159-187 in Kangaroos: Their Ecology and Management in the Sheep Rangelands of Australia. G. Caughley, N. Shepherd, and J. Short, eds. Cambridge Studies in Applied Ecology and Resource Management. Cambridge, U.K.: Cambridge University Press.

Chase, A. 1986. Playing God in Yellowstone: The Destruction of America's First National Park. Boston: Atlantic Monthly Press, 446 pp.

Cheville, N.F., A.E. Jensen, S.M. Halling, F.M. Tatum, D.C. Morfitt, S.G. Hennager, and W.M. Frerichs. 1992. Bacterial survival, lymph node changes, and immunologic responses of cattle vaccinated with standard and mutant strains of Brucella abortus. Am. J. Vet. Res. 53(10):1881-1888.

Cheville, N.F., M.G. Stevens, A.E. Jensen, F.M. Tatum, and S.M. Halling. 1993. Immune responses and protection against infection and abortion in cattle experimentally vaccinated with mutant strains of Brucella abortus. Am. J. Vet. Res. 54(10):1591-1597.

Cheville, N.F., A.E. Jensen, D.C. Morfitt, and T.J. Stabel. 1994. Cutaneous delayed hypersensitivity reactions of cattle vaccinated with mutant strains of Brucella abortus, using brucellins prepared from various brucellar strains. Am. J. Vet. Res. 55(9):1261-1266.

Cheville, N.F., S.C. Olsen, A.E. Jensen, M.G. Stevens, M.V. Palmer, and A.M. Florance. 1996. Effects of age at vaccination on efficacy of Brucella abortus strain RB51 to protect cattle against brucellosis. Am. J. Vet. Res. 57(8):1153-1156.

Church, D.C., G.E. Smith, J.P. Fontenot, and A.T. Ralston. 1971. Digestive physiology and nutrition of ruminants. Volume 2. D.C. Church, ed. Corvallis: Oregon State University.

Clark, W.W., and J.D. Kopec. 1985. Movement of Yellowstone Park Brucellosis Infected and Exposed Bison. U.S. Animal Health Assoc.

Clutton-Brock, T.H., F.E. Guinness, and S.D. Albon. 1982. Red Deer: Behavior and Ecology of Two Sexes. Chicago: University of Chicago Press. 378 pp.

Coburn, D. R. 1948. Special report of field assignment at Yellowstone National Park, January 10-January 29, 1948. Archives, Yellowstone National Park, WY. 30 pp.

Cole, G.F. 1976. Management involving grizzly and black bears in Yellowstone National Park, 1970-75. Nat. Resour., Rep. No. 9 USDI/NPS/Yellowstone National Park. 26 pp.

Cook, D.R., and G.C. Kingston. 1988. Isolation of Brucella suis biotype 1 from a horse. Aust. Vet. J. 65(5):162-163.

Corner, A.H., and R. Connell. 1958. Brucellosis in bison, elk and moose in Elk Island National Park, Alberta, Canada. Can. J. Comp. Med. 22(1):9-21.

Coughenour, M.B., and F.J. Singer. 1996. Elk population processes in Yellowstone National Park under the policy of natural regulation. Ecol. Applic. 6:573-593.

Craighead, J.J., G. Atwell, and B.W. O'Gara. 1972. Elk migration in and near Yellowstone National Park. Wildl. Monogr. 29. 48 pp.

Craighead, J.J., J.S. Sumner, and J.A. Mitchell. 1995. The Grizzly Bears of Yellowstone: Their Ecology in the Yellowstone Ecosystem, 1959-1992. Washington, D. C.: Island Press, 535 pp.

Cronin, M.A. 1991. Mitochondrial-DNA phylogeny of deer (Cervidae). J. Mammal. 72:553-566.

Davies, G., E. Cocks, and N. Hebert. 1980. *Brucella abortus* (strain 19) vaccine: (a) determination of the minimum protective dose in cattle; (b) the effect of vaccinating calves previously inoculated with anti-Brucella abortus serum. J. Biol. Stand. 8(3):165-175.

Davis, D.S., W.J. Boeer, J.P. Mims, F.C. Heck, and L.G. Adams. 1979. *Brucella abortus* in coyotes. I. A serologic and bacteriologic survey in eastern Texas. J. Wildl. Dis. 15(3):367-372.

Davis, D.S., F.C. Heck, J.D. Williams, T.R. Simpson, and L.G. Adams. 1988. Interspecific transmission of *Brucella abortus* from experimentally infected coyotes (*Canis latrans*) to parturient cattle. J. Wildl. Dis. 24(3):533-537.

Davis, D.S., J.W. Templeton, T.A. Ficht, J.D. Williams, J.D. Kopec, and L.G. Adams. 1990. *Brucella abortus* in captive bison. I. Serology, bacteriology, pathogenesis, and transmission to cattle. J. Wildl. Dis. 26(3):360-371.

Davis, D.S., J.W. Templeton, T.A. Ficht, J.D. Huber, R.D. Angus, and L.G. Adams. 1991. *Brucella abortus* in bison. II. Evaluation of strain 19 vaccination of pregnant cows. J. Wildl. Dis. 27(2):258-264.

Davis, D.S., J.W. Templeton, T.A. Ficht, J.D. Williams, J.D. Kopec, and L.G. Adams. 1995. Response to the critique of brucellosis in captive bison. J. Wildl. Dis. 31(1): 111-114.

Denniston, C. 1977. Small population size and genetic diversity: Implications for endangered species. Pp. 281-289 in Endangered Birds: Management Techniques for Preserving Threatened Species. S.A. Temple, ed. Madison: University of Wisconsin Press.

Deyoe, B.L. 1980. Brucellosis. Pp. 285-298 in Bovine Medicine and Surgery, H.E. Amstutz, ed. Santa Barbara: American Veterinary Publications, Inc.

Dieterich, R.A., J.K. Morton, and R.L. Zarnke. 1991. Experimental *Brucella suis* biovar 4 infection in a moose (*Alces alces*). J. Wildl. Dis. 27(3):470-472.

Dobson, A., and M. Meagher. 1996. The population dynamics of brucellosis in the Yellowstone National Park. Ecology 77(4):1026-1036.

Duffield, B.J., T.A. Streeten, and G.A. Spinks. 1984. Isolation of *Brucella abortus* from lymph nodes of cattle from infected herds vaccinated with low dose strain 19. Aust. Vet. J. 61(12):411-412.

Eberhardt, L.L. 1970. Correlation, regression, and density dependence. Ecology 51:306-310.

Eberhardt, L.L., and R. R. Knight. 1996. How many grizzlies in Yellowstone? J. Wildl. Manage. 60:416-421.

Farnes, P.E. 1996. An index of winter severity for elk. Pp. 303-306 in Effects of Grazing by Wild Ungulates in Yellowstone National Park. F. J. Singer, ed. Technical Report NPS/NRYELL/NRTR/96-01, USDI National Park Service, Natural Resource Program Center, Natural Resource Information Division, Denver, CO.

Ferguson, M.A.D. 1997. Rangiferine brucellosis in Baffin Island. J. Wildl. Dis. 33:536-543.

Ferrlicka D. 1992. 1991-1992 Yellowstone Bison Summary. Montana Department of Livestock.

Fisher, R.A. 1930. The genetical theory of natural selection. Oxford: The Clarendon Press. 272 pp.

Fitch, C.P., L.M. Bishop, and W.L. Boyd. 1932. A study of bovine blood, urine and feces for the presence of B. abortus Bang. Proc. Soc. Exp. Biol. Med. 29:555-558.

Flagg, D.E. 1983. A case history of a brucellosis outbreak in a brucellosis free state which originated in bison. Proc. U.S. Animal Health Assoc. 87:171-172.

Forbes, L.B. 1990. *Brucella abortus* infection in 14 farm dogs. J. Am. Vet. Med Assoc. 196(6):911-916.

Forbes, L.B., and S.V. Tessaro. 1993. Transmission of brucellosis from reindeer to cattle. J. Am. Vet. Med. Assoc. 203(2):289-294.

Forbes, L.B., S.V. Tessaro, and W. Lees. 1996. Experimental studies on *Brucella abortus* in moose (Alces alces). J. Wildl. Dis. 32:94-104.

Frankel, O.H., and M.E. Soulé. 1981. Conservation and evolution. Cambridge, England: Cambridge University Press. 327 pp.

Franson, J.C., and B.L. Smith. 1988. Septicemic pasteurellosis in elk (Cervus elaphus) on the United States National Elk Refuge, Wyoming. J. Wildl. Dis. 24(4):715-7.

Fraser, A.F. 1968. Reproductive Behaviour in Ungulates. London; New York: Academic Press, 202 pp.

GAO (General Accounting Office). 1997. Wildlife Management: Issues Concerning the Management of Bison and Elk Herds in Yellowstone National Park. GAO/T-RCED-97-200. Washington, D.C.: U.S. General Accounting Office.

Garner, M.M., D.M. Lambourn, S.J. Jeffries, P.B. Hall, J.C. Rhyan, D.R. Ewalt, L.M. Polzin, and N.F. Cheville. 1997. Evidence of *Brucella* infection in *Parafilaroides* lungworms in a Pacific harbor seal (*Phoca vitulina richardsi*). J. Vet. Diagn. Invest. 9:298-303.

Garretson, M.S. 1938. The American bison. New York Zoological Society, New York. 254 pp.

Gasaway, W.C., R.O. Stephenson, J.L. Davis, P.E.K. Shephard, and O.E. Burris. 1983. Interrelationships of wolves, prey, and man in interior Alaska. Wildl. Monogr. 84. 50 pp.

Gasaway, W.C., R.D. Boertje, D.V. Grangaard, D.G. Kellyhouse, R.O. Stephenson, and D.G. Larsen. 1992. The role of predation in limiting moose at low densities in Alaska and Yukon and implications for conservation. Wildl. Monogr. 120. 59 pp.

Gates, C.C., D.A. Melton, and R. McLeod. 1991. Cattle diseases in northern Canadian bison: a complex management issue. Proceedings Ongules/Ungulates 91:547-550, Institute de Recherche sur les Grands Mammiferes, Toulouse, France, September 2-6, 1991.

Geist, V. 1982. Adaptive behavioral strategies. Pp. 219-277 in Elk of North America: Ecology and Management. J.W. Thomas, and D.E. Toweill, eds. Harrisburg, PA: Wildlife Management Institute/Stackpole Books.

Georgiadis, N.J., P.W. Kat, and H. Oketch. 1990. Allozyme divergence within the Bovidae. Evolution 44:2135-2149.

Gese, E.M., T.E. Stotts, and S. Grothe. 1996a. Interactions between coyotes and red foxes in Yellowstone National Park, Wyoming. J. Mammal. 77:377-382.

Gese, E.M., R.L. Ruff, and R.L. Crabtree. 1996b. Foraging ecology of coyotes (*Canis latrans*): the influence of extrinsic factors and a dominance hierarchy. Can. J. Zool. 74(5):769-783.

Gese, E.M., R.D. Schultz, M.R. Johnson, E.S. Williams, R.L. Crabtree, and R.L. Ruff. 1997. Serological survey for diseases in free-ranging coyotes (*Canis latrans*) in Yellowstone National Park, Wyoming. J. Wildl. Dis. 33(1):47-56.

Gilsdorf M. 1997. Brucellosis in bison - case studies. Wildl. Dis. Conf., Bozeman MT.

GTNP (Grand Teton National Park, National Elk Refuge, Wyoming Game and Fish Department, and Bridger-Teton National Forest). 1993. Agency draft. The Jackson bison herd, long range management plan and environmental assessment. 97 pp.

Gunther, K.A. 1994. Characteristics of black bears and grizzly bears in Yellowstone National Park. Information paper No. BMO-2. Bear Management Office, Yellowstone National Park. 2 pp.

Gunther, K.A., M.J. Biel, HL. Robison, and H.N. Zachary. 1997. Bear Management Offce administrative annual report for calendar year 1996. On file at Yellowstone National Park, WY.

GYIBC (Greater Yellowstone Interagency Brucellosis Committee). 1997. Greater Yellowstone Interagency Brucellosis Committee "White Paper".

Hamilton, W.D. 1971. Geometry for the selfish herd. J. Theor. Biol. 31(2):295-311.

Harrington, F.H. 1981. Urine-marking and caching behavior in the wolf. Behaviour 76:280-288.

Harrington, R., Jr., and G.M. Brown. 1976. Laboratory summary of Brucella isolations and typing: 1975. Am. J. Vet. Res. 37(10):1241-1242.

Harris, L.D. 1984. The Fragmented Forest: Island Biogeography Theory and the Preservation of Biotic Diversity. Chicago: University of Chicago Press, 211 pp.

Hatier, K.G. 1995. Effects of helping behaviors on coyote packs in Yellowstone National Park, Wyoming. M. Sc. thesis, Montana State University, Bozeman. 78 pp.

Herriges, J.D., Jr., E.T. Thorne, and S.L. Anderson. 1991. Vaccination to control brucellosis in free-ranging elk (*Cervus elaphus*) on western Wyoming feedgrounds. Pp. 107-112 in The Biology of Deer, R.D. Brown, ed. New York: Springer-Verlag.

Hobbs, N.T., D.L. Baker, J.E.Ellis, and D.M. Swift. 1979. Composition and quality of elk diets during winter and summer: A preliminary analysis. Pp. 47-53 in North American Elk: Ecology, Behavior, and Management. M.S. Boyce and L.D. Hayden-Wing, eds. Laramie: University of Wyoming Press.

Honess, R.F., and K.B. Winter. 1956. Diseases of wildlife in Wyoming. Bulletin 9, Wyoming Game and Fish Department, Laramie. 279 pp.

Hong, C.B., J.M. Donahue, R.C. Giles, Jr., K.B. Poonacha, P.A. Tuttle, and N.F. Cheville. 1991 *Brucella-abortus*-associated meningitis in aborted bovine fetuses. Vet. Pathol. 28(6):492-496.

Houston, D.B. 1982. The Northern Yellowstone Elk: Ecology and Management. New York: Macmillan Publishing. 474 pp.

Hudson, M., K.N. Child, D.F. Hatler, K.K. Fujino, and K.A. Hodson. 1980. Brucellosis in moose (*Alces alces*). A serological survey in an open range cattle area of North Central British Columbia recently infected with bovine brucellosis. Can. Vet. J. 21(2):47-49.

Janecek, L.L., R.L. Honeycutt, R.M. Adkins, and S.K. Davis. 1996. Mitochondrial gene sequences and the molecular systematics of the artiodactyl subfamily Bovinae. Mol. Phylogenet. Evol. 6(1):107-119.

Jensen, A.E., N.F. Cheville, D.R. Ewalt, J.B. Payeur, and C.O. Thoen. 1995. Application of pulsed-field gel electrophoresis for differentiation of vaccine strain RB51 from field isolates of *Brucella abortus* from cattle, bison, and elk. Am. J. Vet. Res. 56(3):308-312.

Jensen, A.E., D.R. Ewalt, N.F. Cheville, C.O. Thoen, and J.B. Payeur. 1996. Determination of stability of *Brucella abortus* RB51 by use of genomic fingerprint, oxidative metabolism, and colonial morphology and differentiation of strain RB51 from *B. abortus* isolates from bison and elk. J. Clin. Microbiol. 34(3):628-633.

Jimenez de Bagues, M.P., M. Barberan, C.M. Marin, and J.M. Blasco. 1995. The *Brucella abortus* RB51 vaccine does not confer protection against *Brucella ovis* in rams. Vaccine 13:301-304.

Johnson, D.E. 1951. Biology of the elk calf, *Cervus canadensis nelsoni*. J. Wildl. Manage. 15:396-410.

Johnson, M.R. 1992. The disease ecology of brucellosis and tuberculosis in potential relationship to Yellowstone wolf populations. Pp. 5-73 - 5-92 in Wolves for Yellowstone?: A Report to the United States Congress. Volume IV, Research and Analysis.

Jones, R.L., R.I. Tamayo, W. Porath, N. Geissman, L.S. Selby, and G.M. Buening. 1983. A serologic survey of brucellosis in white-tailed deer (*Odocoileus virginianus*) in Missouri. J. Wildl. Dis. 19(4):321-323.

Kay, C.E. 1990. Yellowstone's northern elk herd: a critical evaluation of the "Natural Regulation" paradigm. Ph.D. dissertation, Utah State University, Logan.

Kearley, B. 1996. Report of telephone conversation between Dr. Sam Holland, South Dakota State Veterinarian and Bill Kearley DVM, Division of Animal Industries, Idaho Department of Agriculture. 4 pp.

King, R.O.C. 1940. Brucella infection in the bull: a progress report of mating experiments with naturally infected bulls. Aust. Vet. J. 16:117-119.

Kiok, P., E.G. Grunbaum, W. Letz, W. Uhl, and K. Mieth. 1978. Dogs as a source of reinfection of brucellosis-free cattle herds. Der Hund als Reinfektionsquelle für Brucellosefreie Rinderbestande. Monatsh Veterinarmed 33(18):700-704.

Kirkpatrick, J.F., J.C. McCarthy, D.F. Gudermuth, S.E. Shideler, and B.L. Lasley. 1996. An assessment of the reproductive biology of Yellowstone bison (*Bison bison*) subpopulations using noncapture methods. Can. J. Zool. 74:8-14.

Kufeld, R. C. 1973. Foods eaten by Rocky Mountain elk. J. Range Manage. 26:106-113.

Lambert, G., T.E. Amerault, C.A. Manthei, and E.R. Goode, Jr. 1960. Further studies on the persistence of *Brucella abortus* in cattle. Proc. U. S. Livestock Sanit. Assoc. pp 109-117.

Lambert, G., C.A. Manthei, and B.L. Deyoe. 1963. Studies on *Brucella abortus* infection in bulls. Am. J. Vet. Res. 24(103):1152-1156.

Lancia, R.A., C.E. Braun, M.W. Collopy, R.D. Dueser, J.G. Kie, C.J. Martinka, J.D. Nichols, T.D. Nudds, W.R. Porath, and N.G. Tilghman. 1996. ARM! for the future: adaptive resource management in the wildlife profession. Wildl. Soc. Bull. 24:436-442.

Lent, P.C. 1974. Mother-infant relationships in ungulates. Pp. 14-55 in The Behaviour of Ungulates and its Relation to Management, Vol 1., V. Geist and F. Walther, eds. IUCN Publications new series No. 24. International Union for Conservation and Natural Resources, Morges, Switzerland.

Leopold, A. 1933. Game management. New York: Charles Scribner's Sons. 481 pp.

Leopold, A. 1940. Song of the Gavilan. J. Wildl. Manage. 4:329-331.

Livezey, K.B. 1979. Social behavior of Rocky Mountain elk at the National Bison Range. M.S. thesis, University of Montana, Bozeman. 142 pp.

Long, B., M. Hinschberger, G. Roby, and J. Kimbal. 1980. Gros Ventre cooperative elk study. Final report 1974-79. Wyoming Fish and Game Department, US Forest Service, Jackson, WY.

Lott, D. F. 1974. Sexual and aggressive behavior of adult male American bison (*Bison bison*). Pp. 382-394 in The Behaviour of Ungulates and its Relation to Management, Vol 1., V. Geist and F. Walther, eds. IUCN Publications new series No. 24. International Union for Conservation of Nature and Natural Resources, Morges, Switzerland.

Lott, D.F., and J.C. Galland. 1985. Parturition in American bison: precocity and systematic variation in cow isolation. Z. Tierpsychol. 69(1):66-71.

Lott, D.F., and S.C. Minta. 1983. Random individual association and social group instability in American bison (*Bison bison*). Z. Tierpsychol. 61(2):153-172.

Mackie, R.J. 1985. The elk-deer-livestock triangle. Pp. 51-56 in Western

Elk Management: A Symposium, G. W. Workman, ed. Utah State University, Logan.

MacMillan, A.P., A. Baskerville, P. Hambleton, and M.J. Corbel. 1982. Experimental *Brucella abortus* infection in the horse: observations during the three months following inoculation. Res. Vet. Sci. 33(3):351-359.

Manthei, C.A., and R.W. Carter. 1950. Persistence of *Brucella abortus* infection in cattle. Am. J. Vet. Res. 11(39):173-180.

Manthei, C.A., D.E. DeTray, and E.R. Goode. 1950. *Brucella* infection in bulls and the spread of brucellosis in cattle by artifical insemination. I. Intrauterine injection. Am. Vet. Med. Assoc. Proc. 87th Annu. Meet. Pp. 177-184.

Marcum, C.L. 1979. Summer-fall food habits and forage preferences of a western Montana elk herd. Pp. 54-62 in North American elk: ecology, behavior, and management. M. S. Boyce and L. D. Hayden-Wing, eds. Laramie: University of Wyoming Press.

Marley, S.E., S.E. Knapp, M.C. Rognlie, J.R. Thompson, T.M. Stoppa, S.M. Button, S. Wetzlich, T. Arndt, and A. Craigmill. 1995. Efficacy of ivermectin pour-on against *Ostertagia ostertagi* infection and residues in the American bison, *Bison bison*. J. Wildl. Dis. 31(1):62-65.

McLaren, B.E., and R.O. Peterson. 1994. Wolves, moose, and tree rings on Isle Royale. Science 266(5190):1555-1558.

McCullough, D.R. 1969. The tule elk: Its history, behavior, and ecology. University of California Publications in Zoology, Vol 88. Berkeley and Los Angeles: University of California Press. 209 pp.

McCullough, D.R. 1985. Long range movements of large terrestrial animals. Pp. 444-465 in Migration: Mechanisms and Adaptive Significance. M.A. Rankin, ed. Contributions in Marine Science, Suppl. Vol 27.

McCullough, D.R. 1990. Detecting density dependence: filtering the baby from the bathwater. Trans. 55th N. Amer. Wildl. and Nat. Resour. Conf. 55:534-543.

McCullough, D.R. 1992. Concepts of large herbivore population dynamics. Pp. 967-984 in Wildlife 2001: Populations, D.R. McCullough and R.H. Barrett, eds. London and New York: Elsevier Applied Science.

McCullough, D.R. 1997. Irruptive behavior in ungulates. Pp. 69-98 in The Science of Overabundance: Deer Ecology and Population Management, W.J. McShea, H.B. Underwood, and J.H. Rappole, eds. Washington and London: Smithsonian Institution Press.

McCullough, Y.B. 1980. Niche separation of seven North American ungulates on the National Bison Range, Montana. Ph.D. dissertation, University of Michigan, Ann Arbor. 226 pp.

McDonald, J.N. 1981. North American Bison: Their Classification and Evolution. Berkeley and Los Angeles: University of California Press. 316 pp.

McEwen, A.D., F.W. Priestley, and J.D. Paterson. 1939. An estimate of a suitable infective dose of *B. abortus* for immunisation tests on cattle. J. Comp. Pathol. Therapeut. 52:116-128.

McHugh, T.C. 1958. Social behavior of the American buffalo (*Bison bison bison*). Zoologica, N. Y. 43:1-40.

McKean, W.T. 1949. A search for Bang's disease. in N.D. deer. North Dakota Outdoors 11:10.

Meador, V.P., B.L. Deyoe, and N.F. Cheville. 1989. Effect of nursing on *Brucella abortus* infection of mammary glands of goats. Vet. Pathol. 26(5):369-375.

Meagher, M.M. 1973. The Bison of Yellowstone National Park. National Park Service Scientific Monograph Series Number One. Washington, D.C.: National Park Service. 161 pp.

Meagher, M. 1986. Bison bison. Mammalian species No. 266. The American Society of Mammalogists. 8 pp.

Meagher, M. 1989. Range expansion by bison of Yellowstone National Park. J. Mammal. 70:670-675.

Meagher, M. 1993. Winter recreation-induced changes in bison numbers and distribution in Yellowstone National Park. Unpublished report. Yellowstone National Park files. 48 pp.

Meagher, M., and M.E. Meyer. 1994. On the origin of brucellosis in bison of Yellowstone National Park: a review. Conservation Biol. 8(3):645-653.

Meagher, M., S. Cain, T. Toman, J. Kropp., and D. Bosman. 1997. Bison in the Greater Yellowstone Area: status, distribution, and management. Pp. 47-55 in Brucellosis, Bison, Elk, and Cattle in the Greater Yellowstone Area: Defining the Problem, Exploring Solutions. T. Thorne et al., eds. Wyoming Game and Fish Department, Cheyenne.

Mech, L.D. 1966. The wolves of Isle Royale. Fauna of the National Parks of the United States Fauna Series 7. USDI/National Park Service, Washington, D. C. 210 pp.

Mech, L.D. 1970. The Wolf: the Ecology and Behavior of an Endangered Species. Garden City, N. J.: Natural History Press. 384 pp.

Mech, L.D., R.E. McRoberts, R.O. Peterson, and R.E. Page. 1987. Relationship of deer and moose populations to previous winters' snow. J. Animal Ecol. 56:615-627.

Merrill, E.H., and M.S. Boyce. 1991. Summer range and elk population dynamics in Yellowstone National Park. Pp. 263-273 in The Greater Yellowstone Ecosystem: Redefining America's Wilderness Heritage, R.B.

Keiter and M.S. Boyce, eds. New Haven, CT: Yale University Press.

Meyer, M.E., and M. Meagher. 1995. Brucellosis in free-ranging bison (*Bison bison*) in Yellowstone, Grand Teton, and Wood Buffalo National Parks: a review. J. Wildl. Dis. 31(4):579-598.

Meyer, M.E., and M. Meagher. 1997. *Brucella abortus* infection in the free-ranging bison of Yellowstone National Park. Pp. 20-32 in Brucellosis, Bison, Elk, and Cattle in the Greater Yellowstone Area: Defining the Problem, Exploring Solutions, E. T. Thorne et al., eds. Wyoming Game and Fish Department, Cheyenne.

Miller, L.A., B.E. Johns, D.J. Elias, and K.A. Crane. 1997. Comparative efficacy of two immunocontraceptive vaccines. Vaccine 15(17-18):1858-62.

Miller, L.H., M.F. Good, and G. Milon. 1994. Malaria pathogenesis. Science 264(5167):1878-1883.

Mitscherlich, E., and E.H. Marth. 1984. Microbial survival in the environment: Bacteria and rickettsiae important in human and animal health. New York: Springer Verlag. 802 pp.

Mohler, J.R. 1917. Abortion disease. In: Annual Reports, U.S. Department of Agriculture, Washington, D.C. pp. 105-106.

Montana Department of Fish, Wildlife, and Parks; USDI/NPS/Yellowstone National Park; USDA/USFS/Gallatin National Forest. 1990. Yellowstone bison: background and issues. 28 pp.

Moore, C.G., and P.R. Schnurrenberger. 1981. A review of naturally occurring *Brucella abortus* infections in wild mammals. J. Am. Vet. Med. Assoc. 179(11):1105-1112.

Morgan, W.J.B., and A. McDiarmid. 1960. The excretion of *Brucella abortus* in the milk of experimentally infected cattle. Res. Vet. Sci. 1:53-5.

Morton, J.K. 1989. *Brucella suis* Type 4 in foxes and their role as reservoirs/vectors among reindeer. Ph. D. dissertation, University of Alaska, Fairbanks. 263 pp.

Morton, J.K., E.T. Thorne, and G.M. Thomas. 1981. Brucellosis in elk. III. Serologic evaluation. J. Wildl. Dis. 17(1):23-31.

Murie, A. 1944. The wolves of Mount McKinley. Fauna of the National Parks of the United States. Fauna Series No. 5. 238 pp.

Murie, O.J. 1951. The Elk of North America. Wildlife Management Institute/Stackpole Books, Harrisburg, PA. 376 pp.

Nicoletti, P.L. 1977. Adult vaccination (Eliminating brucellosis from some cattle herds). Pp. 201-208 in Bovine Brucellosis, an International Symposium, 1976. R.P. Crawford, and R.J. Hidalgo, eds. College Station: Texas A&M University Press.

Nicoletti, P. 1981. The efficacy of strain 19 vaccination in reducing brucellosis in large dairy herds. Calif. Vet. 35(9):35-36.

NRC (National Research Council) 1997. Wolves, bears, and their prey in Alaska: Biological and social challenges in wildlife management. Washington, D.C.: National Academy Press. 207pp.

Oakley, C.A. 1975. Elk distribution in relation to a deferred grazing system. J. Range Manage. 28:274(abstract).

Oldemeyer, J.L., R.L. Robbins, and B.L. Smith. 1990. Effect of feeding level on elk weights and reproductive success at the National Elk Refuge. Proceedings of the 1990 Western States and Provinces Elk Workshop.

Olsen, S.C., N.F. Cheville, R.A. Kunkle, M.V. Palmer, and A.E. Jensen. 1997. Bacterial survival, lymph node pathology, and serological responses of bison (*Bison bison*) vaccinated with *Brucella abortus* strain RB51 or strain 19. J. Wildl. Dis. 33(1):146-151.

Olsen, S.C., A.E. Jensen, M.V. Palmer, and M.G. Stevens. 1998. Evaluation of serologic responses, lymphocyte proliferative responses and clearance from lymphatic organs after vaccination of bison with *Brucella abortus* strain RB51. Am. J. Vet. Res. 59(4):410-415.

Pac, H.I., and K. Frey. 1991. Some population characteristics of the northern Yellowstone Bison herd during the winter of 1988-89. Mont. Dept. Fish, Wildlife, and Parks, Bozeman. 29 pp.

Palmer, M.V., S.C. Olsen, M.J. Gilsdorf, L.M. Philo, P.R. Clarke, and N.F. Cheville. 1996. Abortion and placentitis in pregnant bison (*Bison bison*) induced by the vaccine candidate, *Brucella abortus* strain RB51. Am. J. Vet. Res. 57(11):1604-1607.

Palmer, M.V., and N.F. Cheville. 1997. Effects of oral or intravenous inoculation with *Brucella abortus* strain RB51 vaccine in Beagles. Am. J. Vet. Res. 58:851-856, 1997.

Parker Land and Cattle Company, Inc. vs United States of America (No. 91-CV-0039-B) and Lyle E. Peck vs United States of America (No. 91-CV-0091-B). 1992. U. S. District Court for the District of Wyoming.

Peacock, D. 1997. The Yellowstone massacre. Audubon 99(3) (May-June):40-49, 102-103,06-110.

Peterson, M.J. 1991. Wildlife parasitism, science, and management policy. J. Wildl. Manage. 55:782-789.

Peterson, R.O., and R.E. Page. 1988. The rise and fall of Isle Royale Wolves, 1975-1986. J. Mammal. 69:89-99.

Peterson, M.J., W.E. Grant, and D.S. Davis. 1991a. Bison-brucellosis management: simulation of alternative strategies. J. Wildl. Manage. 55:205-213.

Peterson, M.J., W.E. Grant, and D.S. Davis. 1991b. Simulation of host-parasite interactions within a resource management framework: impact of brucellosis on bison population dynamics. Ecol. Model. 54:299-320.

Pickett, S.T.A., and P.S. White, eds. 1985. The ecology of natural disturbance and patch dynamics. New York: Academic Press. 472pp.

Pickett, S.T.A., V.T. Parker, and P.L. Fiedler. 1992. The new paradigm in ecology: Implications for conservation biology above the species level. Pp. 65-88 in Conservation Biology: The Theory and Practice of Nature Conservation, Preservation, and Management. New York and London: Chapman and Hall.

Poppe, K. 1929. Der infektiöse Abortus des Rindes (Bang-Infektion). In: Handbuch der pathogenen Mikroorganismen, Kolle et al (eds.) Gustav Fischer, Jena. Pp. 693-750.

Pulliam, H.R. 1988. Sources, sinks, and population regulation. American Naturalist 132:652-661.

Rankin, J.E.F. 1965. *Brucella Abortus* in bulls: a study of twelve naturally-infected cases. Vet. Rec. 77:132-135.

Reice, S.R. 1994. Nonequilibrium determinants of biological community structure. Am. Scientist 82:424-435.

Remenëtïsova, M.M. 1987. Brucellosis in wild animals. (*Bruëtisellez dikikh zhivotnykh.* 1985). I.A. Galuzo and E.V. Gvozdev, eds. Amerind Publishing Co. Pvt. Ltd., New Delhi. 323 pp.

Rhyan, J.C., D.A. Saari, E.S. Williams, M.W. Miller, A.J. Davis, and A.J. Wilson. 1992. Gross and microscopic lesions of naturally occurring tuberculosis in a captive herd of wapiti (*Cervus elaphus nelsoni*) in Colorado. J. Vet. Diagn. Invest. 4(4):428-433.

Rhyan, J.C., W.J. Quinn, L.S. Stackhouse, J.J. Henderson, D.R. Ewalt, J.B. Payeur, M. Johnson, and M. Meagher. 1994. Abortion caused by *Brucella abortus* biovar 1 in a free-ranging bison (*Bison bison*) from Yellowstone National Park. J. Wildl. Dis. 30(3):445-446.

Rhyan, J.C., K. Aune, D.R. Ewalt, J. Marquardt, J.W. Mertins, J.B. Payeur, D.A. Saari, P. Schladweiler, E.J. Sheehan, and D. Worley. 1997. Survey of free-ranging elk from Wyoming and Montana for selected pathogens. J. Wildl. Dis. 33:290-298.

Rhyan, J.C., S.D. Holland, T. Gidlewski, D.A. Saari, A.E. Jensen, D.R. Ewalt, D.G. Hennager, S.C. Olsen, and N.F. Cheville. 1998. Seminal vesiculitis and orchitis caused by *Brucella abortus* biovar 1 in young bison bulls from South Dakota. (Submitted).

Rinehart, J.E., and L.D.Fay. 1981. Brucellosis. Pp. 148-154 in Diseases and Parasites of White-Tailed Deer. W.R. Davidson, F.A. Hayes, V.F. Nettles, and F.E. Kellogg, eds. Athens, Ga.: Southeastern Cooperative

Wildlife Disease Study, Department of Parsitology, College of Veterinary Medicine, University of Georgia. 458 pp.

Robison, C.D. 1994. Conservation of germ plasm from *Brucella abortus*-infected bison (*Bison bison*) using natural service, embryo transfer, and *in vitro* maturation/*in vitro* fertilization. MS thesis, Texas A & M University, College Station. 34 pp.

Roe, F.G. 1951. The North American Buffalo: a Critical Study of the Species in its Wild State. Toronto, Canada: University of Toronto Press, 991 pp.

Roffe, T.J., K. Aune, J.C. Rhyan, L.M. Philo, M. Gilsdorf, S.C. Olsen, T. Gidlewski, and D.R. Ewalt. 1997. Epidemiology and pathogenesis of brucellosis in Yellowstone National Park bison (*Bison bison*). (Abstract) In Proc. Intl. Symp. on Bison Ecology and Management in North America, June 4-7, 1997, Bozeman, MT.

Rogers, E.B. 1950. Bison reports. Lamar bison herd reduction. Memo. to the Regional Director, 28 March, 1950. Archives, Yellowstone National Park, WY. 6 pp.

Rush, W.M. 1932. Bang's disease in Yellowstone National Park buffalo and elk herds. J. Mammal. 13:371-372.

Russell, E.W.B. 1994. Land use history. (Review of a symposium on land use history and ecosystem processes: Inexorably connected). Bull. of Ecol. Soc. Amer. 74:35-36.

Sandford, S.E. 1995. Outbreaks of yersiniosis caused by Yersinia pseudotuberculosis in farmed cervids. J. Vet. Diagn. Invest. 7(1):78-81.

Schullery, P. 1992. The Bears of Yellowstone. Worland, WY: High Plains Publishing . 318 pp.

Shotts, E.B., W.E. Greer, and F.A. Hayes. 1958. A preliminary survey of the incidence of brucellosis and leptospirosis among white-tailed deer. (*Odocoileus virginianus*) of the Southeast. J. Am. Vet. Med. Assoc. 133(7):359-361.

Simberloff, D. 1982. A succession of paradigms in ecology: Essentialism to materialism and probabilism. Pp. 63-99 in Conceptual Issues in Ecology. E. Saarinen, ed. Boston: D. Reidel Publishing Co.

Sinclair, A.R.E. 1983. Management of conservation areas as ecological baseline controls. Pp. 13-22 in Management of large mammals in African conservation areas. R. N. Owen-Smith, ed. Pretoria, R.S.A.: Haum Educational Publishers.

Singer, F.J., and J.E. Norland. 1996. Niche relationships within a guild of ungulate species in Yellowstone National Park, Wyoming, following release from artificial controls. Pp. 345-360 in F. J. Singer, ed. Effects of grazing by wild ungulates in Yellowstone National Park. Technical report NPS/NRYELL/NRTR/96-01, USDI National Park Service, Natural Resource Program Center, Natural Resource Information Division, Denver, CO.

Skovlin, J.M., P.J. Edgerton, and R.W. Harris. 1968. The influence of cattle management on deer and elk. Trans. N. Amer. Wildl. Conf. 33:169-81.

Smith, B.L. 1994. Out-of-season births of elk calves in Wyoming. Prairie Natural. 26:131-136.

Smith, B.L., and R.L. Robbins. 1994. Migrations and management of the Jackson elk herd. Washington, D.C.: U.S. Department of the Interior, National Biological Survey. Resource Pub. No. 199. 61 pp.

Smith B.L., and T. Roffe. 1992. Diseases among elk of the Yellowstone ecosystem, USA. Proc. 3rd Intern. Wildl. Ranching Symp. Pretoria, South Africa, October 1992.

Smith, B., and T. Roffe. 1997. Evaluation of studies of Strain 19 *Brucella abortus* vaccine in elk: clinical trials and field applications. Unpublished manuscript.

Smith, S., E.T. Thorne, S. Anderson-Pistono, and T.J. Kreeger. 1995. Efficacy of brucellosis vaccination of free-ranging elk in the Greater Yellowstone Area: The first 10 years. Proc 45th Ann. Wildl. Dis. Assoc. Conf. p. 20.

Smith, S.G., S. Kilpatrick, A.D. Reese, B.L. Smith, T. Lemke, and D. Hunter. 1997. Wildlife habitat, feedgrounds, and brucellosis in the Greater Yellowstone Area. Pp. 65-76 in E.T. Thorne, M.S. Boyce, P. Nicoletti, and J.J. Kreeger, eds. Brucellosis in bison, elk and cattle in the Greater Yellowstone Area: Defining the problem, exploring solutions. Cheyenne, WY: Wyoming Fish and Game Department.

Sousa, W.P. 1984. The role of disturbance in natural communities. Annual Rev. Ecol. and System. 15:333-341.

Spraker, T.R., M.W. Miller, E.S. Williams, D.M. Getzy, W.J. Adrian, G.G. Schoonveld, R.A. Spowart, K.I. O'Rourke, J.M. Miller, and P.A. Merz. 1997. Spongiform encephalopathy in free-ranging mule deer (*Odocoileus hemionus*), white-tailed deer (*Odocoileus virginianus*) and Rocky Mountain elk (*Cervus elaphus nelsoni*) in northcentral Colorado. J. Wildl. Dis. 33(1):1-6.

SPSS Inc. 1996. Systat 6.0 for Windows: Statistics. SPSS Inc, Chicago. Englewood Cliffs, N.J.: Prentice Hall. 751 pp.

Steen, M.O., H. Brohn, and D. Robb. 1955. A survey of brucellosis in white-tailed deer in Missouri. J. Wildl. Manage. 19:320-321.

Stormont, C. 1993. Dr. Stormont's most recent gene frequency charts and tables. Page 65 in Bison Breeder's Handbook, third edition. H. Danz, ed. Denver, CO: American Bison Association. 112 pp.

Taber, R.D., K. Raedeke, and D.A. McCaughran. 1982. Population characteristics. Pp. 279-298 in Elk of North America: Ecology and Management. J.W. Thomas and D.E. Toweill, eds. Harrisburg, PA: Wildlife Management Institute/Stackpole Books.

Taylor, S. K. 1992. Free-ranging bison and brucellosis in Yellowstone National Park. Pp. 95-98 in Proc. Joint Conf. Amer. Assoc. Zoo Vet. and the Amer. Assoc. Wildl. Vet., Nov. 15-19, 1992, Oakland, CA.

Taylor, S.K., V.M. Lane, D. Hunter, K.G. Eyre, S. Kaufman, S. Frye, and M.R. Johnson. 1997. Serologic survey for infectious pathogens in free-ranging American bison. J. Wildl. Dis. 33:308-311.

Telfer, E.S., and A. Cairns. 1979. Bison-wapiti interrelationships in Elk Island National Park, Alberta. Pp. 114-121 in North American elk: Ecology, Behavior, and Management. M. S. Boyce and L.D. Hayden-Wing, eds. Laramie: University of Wyoming.

Tessaro S.V. 1986. The existing and potential importance of brucellosis and tuberculosis in Canadian wildlife: a review. Can. Vet. J. Rev. Vet.Can. 27(3):119-124.

Tessaro, S.V. 1987. A descriptive and epidemiologic study of brucellosis and tuberculosis in bison in northern Canada. Ph.D. dissertation, University of Saskatchewan, Saskatoon. 320 pp.

Tessaro, S.V. 1989. Review of the diseases, parasites and miscellaneous pathological conditions in North American bison. Can. Vet. J. 30(5):416-422.

Tessaro, S.V., L.B. Forbes, and C. Turcotte. 1990. A survey of brucellosis and tuberculosis in and around Wood Buffalo National Park, Canada. Can. Vet. J. 31(3):174-180.

Thoen, C.O., W.J. Quinn, L.D. Miller, L.L. Stackhouse, B.F. Newcomb, and J.M. Ferrell. 1992. *Mycobacterium bovis* infection in North American elk (*Cervus elaphus*). J. Vet. Diagn. Invest. 4(4):423-427.

Thomsen, A. 1943. Does the bull spread infectious abortion in cattle? Experimental studies from 1936 to 1942. J. Comp. Pathol. Therap. 53:199-211.

Thorne E.T., and J.D. Herriges.Jr. 1992. Brucellosis, wildlife and conflicts in the Greater Yellowstone Area. Pp. 453-465 in Trans. 57th N. Amer. Wildl. Nat. Resour. Conf., Washington, D.C.

Thorne, E.T., J.K. Morton, and W.C. Ray. 1979. Brucellosis, its affect and impact on elk in western Wyoming. Pp. 212-220 in North American Elk: Ecology, Behavior, and Management. M.S. Boyce, and L.D. Hayden-Wing, eds. Laramie: University of Wyoming.

Thorne, E.T., and J.K. Morton. 1978. Brucelosis in elk. II. Clinical effects and means of transmission as determined through artifical infection. J. Wildl. Dis. 14(3):280-291.

Thorne, E.T., J.D. Herriges, and A.D. Reese. 1991. Bovine brucellosis in elk: Conflicts in the Greater Yellowstone area. Pp. 296-303 in Proc. Elk Vulnerability Symp., A.G. Christensen, L.J. Lyon, and T.N. Lonner, eds., Montana State University, Bozeman.

Thorne, E.T., S.G. Smith, K. Aune, D. Hunter, and T.J. Roffe. 1997. Brucellosis: the disease in elk. Pp. 33-44 in Brucellosis, Bison, Elk, and Cattle in the Greater Yellowstone Area: Defining the Problem, Exploring Solutions, E.T. Thorne et al., eds. Wyoming Game and Fish Department, Cheyenne.

Thorne, E.T., T.J. Walthall, and H.A. Dawson. 1981. Vaccination of elk with strain 19 *Brucella abortus*. Proc. U. S. Animal Health Assoc. 85:359-374.

Thorpe, B.D., R.W. Sidwell, J.B. Bushman, K.L. Smart, and R. Moyes. 1965. Brucellosis in wildlife and livestock of west central Utah. J. Am. Vet. Med Assoc. 146(3):225-232.

Toman, T.T., T. Lemke, L. Kuck, B.L. Smith, S.C. Smith, and K. Aune. 1997. Elk in the Greater Yellowstone Area: status and management. Pp. 56-64 in Brucellosis, Bison, Elk, and Cattle in the Greater Yellowstone Area: Defining the Problem, Exploring Solutions. E.T. Thorne et al., eds.

Toshach S. 1963. Brucellosis in the Canadian arctic. Can. J. Pub. Health 54(6):271-275.

Trainer, D.O., and R.P, Hanson. 1960. Leptospirosis and brucellosis serological reactors in Wisconsin deer, 1957-1958. J. Wildl. Manage. 24:44-52.

Tunnicliff, E.A., and H. Marsh. 1935. Bang's disease in bison and elk in the Yellowstone National Park and on the National Bison Range. J. Am. Vet. Med. Assoc. 86:745-752.

Turner, J.W., Jr., J.F. Kirkpatrick, and I.K.M. Liu. 1996a. Effectiveness, reversibility, and serum antibody titers associated with immunocontraception in captive white-tailed deer. J. Wildl. Manage. 60(1):45-51.

Turner, J.W., I.K. Liu, and J.F. Kirkpatrick. 1996b. Remotely delivered immunocontraception in free-roaming feral burros (*Equus asinus*). J. Reprod. Fertil. 107(1):31-35.

USDA (United States Department of Agriculture). Animal and Plant Health Inspection Service, Veterinary Services. 1984. Brucellosis Eradication: Uniform Methods and Rules. U. S. Government Printing Office, Washington D.C.

Van Camp, J., and G.W. Calef. 1987. Population dynamics of bison. Pp. 21-24 in Bison Ecology in Relation to Agricultural Development in the Slave River Lowlands, N.W.T. Canadian Wildlife Service.

Van Vuren, D. 1982. Comparative ecology of bison and cattle in the Henry Mountains, Utah, summer diets, preferred forages, distributions, ranges. Pp. 449-457 in Proceedings of the Wildlife-Livestock Relationships Symposium, Coeur d'Alene, ID, 20-22 April, 1981. J.M. Peeke, and P.D. Dalke, eds. Moscow, Idaho: Forest, Wildlife, and Range Experiment Station.

Van Vuren, D. 1983. Group dynamics and summer home range of bison in southern Utah. J. Mammal. 64(2):329-332.

Vander Wall, S.B. 1990. Food Hoarding in Animals. Chicago: University of Chicago Press. 445 pp.

Walters, C.J. 1986. Adaptive management of renewable resources. New York: Macmillan. 374 pp.

Weaver, J.L. 1977. Coyote-food base relationships in Jackson Hole, Wyoming. M. S. thesis, Utah State University, Logan.

Weaver, J.L. 1978. The wolves of Yellowstone. Natural Resources Report Number 14. USDI National Park Service, Washington, D.C. 38 pp.

Western, D., and R.M. Wright. 1994. Natural Connections: Perspectives in Community-based Conservation. Washington, D.C.: Island Press. 581 pp.

White, G.C., and R.M. Bartmann. 1997. Density dependence in deer populations. Pp. 120-135 in W.J. McShea, H.B. Underwood, and J.H. Rappole, eds. The science of overabundance: deer ecology and population management. Washington and London: Smithsonian Institution Press.

White, G.C., and P.P. Swett. 1935. Bang's disease infection transmitted to a dairy herd by horses. J. Am. Vet. Med. Assoc. 87:146-150.

Whitlock, C. 1993. Postglacial vegetation and climate of Grand Teton and southern Yellowstone National Parks. Ecol. Monogr. 63:173-198.

Williams, E.S., S.P. Snyder, and K.L. Martin. 1983. Experimental infection of some North American wild ruminants and domestic sheep with *Mycobacterium paratuberculosis*: clinical and bacteriological findings. J. Wildl. Dis. 19(3):185-191.

Williams, E.S., E.T. Thorne, S.L. Anderson, and J.D. Herriges, Jr. 1993. Brucellosis in free-ranging bison (*Bison bison*) from Teton County, Wyoming. J. Wildl. Dis. 29(1):118-122.

Williams, E.S., S.L. Cain, and D.S. Davis. 1997. Brucellosis: the disease in bison. Pp. 7-19 in E. T. Thorne et al., eds., Brucellosis, bison, elk, and cattle in the Greater Yellowstone Area: Defining the problem, exploring solutions. Wyoming Game and Fish Department, Cheyenne.

Wilson, W. 1969. Problems in the speciation of American fossil bison. *In* Post-Pleistocene man and his environment on the Northern Plains. Proceedings of the First Annual Paleo-environmental Workshop. Student's Press, University of Calgary, Calgary.

Xie, X. 1986. Orally administered brucellosis vaccine: *Brucella suis* strain 2 vaccine. Vaccine 4:212-216.

Yellowstone National Park. 1997. Yellowstone's northern range: complexity and change in a wildland ecosystem. National Park Service, Mammoth Hot Springs, WY. 148 pp.

Yellowstone Science. 1995. News & notes. Yellowstone Science 3(1) (Winter 1995):17-18.

Yellowstone Science. 1997. News & notes. Yellowstone Science 5(2) (Spring 1997):21.

Zarnke, R.L. 1983. Serologic survey for selected microbial pathogens in Alaskan wildlife. J. Wildl. Dis. 19(4):324-329.

Zaugg, J.L., S.K. Taylor, B.C. Anderson, D.L. Hunter, J. Ryder, and M. Divine. 1993. Hematologic, serologic values, histopathologic and fecal evaluations of bison from Yellowstone National Park. J. Wildl. Dis. 29(3):453-457.

Appendix A
Questions Addressed by the NRC Study

I. What are the factors that determine the risk of transmission of Brucella abortus to cattle from bison in Yellowstone National Park and Grand Teton National Park?

 a. What is the state of scientific understanding of transmission of Brucella abortus between wildlife species and between wildlife and cattle?

 b. Does Brucella abortus affect the reproductive potential in bison generally, and specifically bison in the GYA?

 c. Does Brucella abortus pose a risk of transmission when it occurs in bison but is not present in the reproductive system? What risk is associated with infected males? (Is it dynamic?)

 d. What is the relationship among serology, culture test results, and likelihood of infectiousness?

 e. What is true prevalence of Brucella abortus in GYA bison and elk? What information is available regarding the prevalence of Brucella abortus in GYA bison and elk? What information is available regarding the prevalence of Brucella abortus in other mammals in the GYA?

 f. What is the risk of direct or indirect (via aborted fetus, placenta, body fluids deposited on the ground, etc.) transmission of Brucella abortus from bison to cattle, from elk to cattle, and from elk to bison or vice versa? What is known about the prevalence of Brucella abortus in GYA wildlife other than bison or elk and risk of transmission to cattle?

 g. In the event that Brucella abortus is removed from bison but not simultaneously from elk, what is the risk that elk will serve as a reinfection pathway for bison?

 h. What is the known risk of Brucella abortus transmission compared with other disease?

i. What is the state of scientific understanding of the safety and effectiveness of existing vaccines to control brucellosis?

j. Why are these vaccines less effective in bison than in cattle?

k. If a vaccination program specific to bison were undertaken, would the outcome have a high likelihood of success given the presence of Brucella abortus in elk and other wildlife?

II. Based solely on scientific considerations, what is known about the relative risk reduction potentials of the various optional approaches to reducing the risk of transmission of Brucella abortus from wildlife to cattle?

- Vaccinating bison
- Vaccinating cattle
- Separating cattle and bison during the bison abortion season or through the entire bison birthing season
- Limiting cattle on the proximity of the park borders to steers only

III. What is the role of vaccine development for bison and elk?

a. Can Brucella abortus be eliminated totally from the GYA by development and use of a vaccine?

b. What would be the theoretical tradeoffs between a vaccine-only approach and a vaccination approach combined with a test and slaughter program?

Appendix B

Bison in the Greater Yellowstone Area
Draft Agenda
24-25 July 1997
Room 108, Reid Hall
Bozeman, MT

---

24 July 1997

8:45

    a.   Opening remarks, introductions
            Lee Paulson, Project Director
            Norman Cheville, Principal Investigator
            Dale McCullough, Principal Investigator

    b. Comments
            Dan Huff, National Park Service, Department of the Interior
            Jack Rhyan, APHIS, U. S. Department of Agriculture
            Bob Hillman, GYBIC

Factors in transmission

    c. Serology and infection; epidemiology and pathogenesis
            Tom Roffe
    d. *Brucella abortus and* reproductive tissues
            Jack Rhyan

e. Risk of transmission
Paul Nicoletti

10:15 Break

10:45 f. Relationship among serology, culture test results, and likelihood of infectiousness
Tom Roffe
g. Genetic diversity and disease resistance in bison with active disease eradication
Joe Templeton
h. Molecular biology and *Brucella abortus*
Peter Gogan

12:15 Lunch break

1:30 i. Modelling
Mike Miller
k. Population changes and distribution
Mary Meagher

3:15 Break

3:45 m. Population dynamics, preliminary data
Peter Gogan
n. Issues in vaccination
Fred Enright
Phil Elzer
o. Safety and effectiveness of existing vaccines
Steve Olsen

5:00 Adjourn

25 June 1997

8:45 p. Elk as a reinfection pathway for bison
Terry Kreeger
q. Outcomes for vaccination program specific to bison given the presence of *Brucella abortus* in elk and other wildlife
Terry Kreeger

10:15   Break

10:45   r. Role of vaccine development for bison and elk

Can *Brucella abortus* be eliminated totally from the GYA by
development and use of a vaccine?
What would be the theoretical tradeoffs between a vaccine-only
approach and a vaccination approach combined with a test and
slaughter program?

Steve Olsen
Fred Enright
Phil Elzer

11:45   Discussion

12:15   Lunch break

1:30   Public comments from interested parties

3:30 Adjourn

Brucellosis in the Greater Yellowstone Area
Agenda
4 August 1997
National Museum of Wildlife Art
Jackson, Wyoming

8:45   Opening remarks, introductions
         Lee Paulson, Project Director
         Norman Cheville, Principal Investigator
         Dale McCullough, Principal Investigator

         Comments
         Bob Schiller, National Park Service, Department of the Interior
         Jack Rhyan, APHIS, U.S. Department of Agriculture
         Art Reese, Wyoming Game and Fish Department

9:45   Molecular genetics
         Betsy Bricker, National Animal Disease Center

10:15   Break

10:45   Experience in Montana
         Keith Aune, Montana Department of Fish, Wildlife and Parks

         Issues in transmission
         Beth Williams, University of Wyoming

         Brucellosis and wildlife research in Yellowstone and Grand Teton
         National Parks
         Wayne Brewster, National Park Service

12:00   Lunch

1:00   Research in elk
         Terry Kreeger, Wyoming Game and Fish

         RB51 in elk
         Phil Elzer, Louisiana State University

Vaccine applications at feedgrounds; habitat improvement
Scott Smith, Wyoming Game and Fish

2:30   Break

3:00   Public comment

5:00   Adjourn

EVALUATION OF *BRUCELLA ABORTUS* VACCINE STRAIN RB51 IN BISON

Philip H. Elzer[1] and Donald S. Davis[2]

[1]Department of Veterinary Science, Louisiana State University Agricultural Center, Baton Rouge, LA 70803 and [2]Department of Veterinary Pathobiology, Texas A&M University, College Station, TX 77843.

Introduction:

Host
Bison - American Buffalo, *Bison bison*
Organism
Brucella abortus
> first isolated in 1930's from the testicle of bull on National Bison Range, Moiese, MT

> Serological positive animals found in 1917 in Yellowstone National Park
Disease
Reproductive disease that causes abortions (late term) that have been documented in the wild.
Problem
Bison which inhabit the Greater Yellowstone area
> approximately 4000 animals with up to 50% seropositive for brucellosis

Cattle grazing in areas adjacent to the park may be susceptible to infection - Wyoming, Montana, and Idaho.

These states could lose their brucellosis-free status.
Vaccine strain
Brucella abortus RB51 - rough derivative of virulent *Brucella abortus* strain 2308. Multiple passages on Rifampin led to the loss of the O-polysaccharide side chain of the LPS. Therefore vaccination with this strain does not lead to the production of antibodies which will interfere with sero-diagnostic test for brucellosis. RB51 provides protection against virulent challenge with strain 2308 in a variety of species including cattle, goats, swine, elk and mice. In cattle, RB51 produces similar protection to that achieved with vaccination with S19 without vaccinal titers. In Pregnant animals, RB51 has also been found to be less pathogenic than S19 in that it induces fewer abortions.

The purpose of this experiment was to evaluate the safety and pathogenesis of *Brucella abortus* strain RB51 in adult and young bison from a previously exposed herd.

Materials and Methods:

Animals

North American buffalo (*Bison bison*) were obtained from a reactor herd in Kansas. The herd contained 3 reactor animal as measured by conventional brucellosis serology. The herd was composed of 10 adult males, 7 calves and 14 adult females. The animals were shipped to and housed at Texas A&M University, College Station, TX throughout the experiment.

Vaccine

Brucella abortus Strain RB51 was obtained from Colorado Serum Co. and rehydrated according to the manufacturer's instructions.

Dose

Adult males and calves received $1\text{-}3\times10^{10}$ colony forming units subcutaneously (standard calfhood dose in cattle). Adult females received 1×10^9 colony forming units subcutaneously (standard adult dose in cattle).

Experimental Design

The adult males and calves plus 5 non-pregnant cows were divided into 2 groups; group 1 was slaughtered at 13 weeks post vaccination, and group 2 was slaughtered at 16 weeks post vaccination.

Pregnant females were monitored until parturition, and delivery status was recorded. Live calves remained with the cows, and dead or weak calves were cultured for *Brucella*.

Tissue collection

The following tissues were collected aseptically at slaughter: liver, spleen, various lymph nodes, and reproductive tracts.

Bacterial culture

All of the tissues were homogenized in sterile distilled water, and the homogenates were plated on *Brucella* selective media.

Serology

Standard western blot analysis using RB51 and smooth field strain cell lysates were performed on all pre- and post-vaccination serum samples.

Results

Culture data

Table 1. Culture data from adult males, non-pregnant females and calves at 13 and 16 weeks post vaccination with RB51.

Weeks post-vaccination	*Brucella abortus* strain recovered	
	Field strain*	RB51
13	1 adult male+	none
16	none	none

*rifampin sensitive, smooth organisim
+animal number 3

Results

Serology Data

Table 2. Western blot analysis of serum samples taken before vaccination and at slaughter using cell lysates from *Brucella abortus* smooth strain 2308 and rough strain RB51.

Animal number	Pre-vaccination		Post-vaccination	
	2308	RB51	2308	RB51
1	−	−	−	+ + +
3*	+ + +	+	+ + +	+ +
5	+ + +	+	+ + +	+ +
8	−	+ / −	+	+ +
9	−	+ / −	+	+ +
11	−	−	−	+ +
13	+ + + +	+ +	+ + + +	+ + +
16	+ / −	−	+ / −	+ + +
19	−	+	−	+ + +
20	−	+ / −	−	+ + +
21	−	−	−	+ + +
22	−	+ / −	−	+ + +
24	−	+ / −	−	+ +
25	−	−	−	+ + + +
26	−	−	−	+ +
28	−	−	−	+ + +
29	−	+	−	+ + +
30	−	+ +	−	+
31	−	−	−	+ / −

*culture positive animal (field strain)

Results

Fetal Pathogenesis

Table 3. Delivery status of female bison vaccinated with 1×10^9 colony forming units of RB51.

No. animals	Abortion	Live birth	Dystocia	Pending
9	0	5	1*	3

* culture negative for RB51 and field strain

Note: another pregnant animal was necropsied at 16 weeks post vaccination, and both the cow and the calf (150 days) were culture negative for RB51 and field strain.

Future Studies:

Determination of the vaccine efficacy of strain RB51 in female bison.

Group 1. Controls - saline subcutaneously
Group 2. RB51 vaccinates (this study) + another vaccination this year
Group 3. RB51 vaccinates (Idaho) + another vaccination this year
Group 4.   RB51 vaccinates (this year)

All of the animals will be vaccinated in September.
The animals will be bred between October and November.
Animals will be challenged in the conjunctival sac with *Brucella abortus* strain 2308 (1×10^7 colony forming units). Delivery and culture status will be monitored.

Conclusions:

1. Vaccination with RB51 in adult or young bison does not result in sero-conversion on standard brucellosis diagnostic tests.
2. Vaccination with RB51 does not result in any gross pathological lesions in calves or adult males.
3. RB51 does not appear to be pathogenic in adult males, non-pregnant females, or calves as measured by increased or prolonged colonization.

4. RB51 does not appear to be pathogenic to adult pregnant females when administered to animals from a reactor herd.

5. Further studies are necessary to determine the vaccine efficacy in bison.

Acknowledgments:

APHIS Brucellosis Research Committee
Dr. Jack Rhyan - NVSL
Dr. Mike Gilsdorf - USDA
Dr. Steve Olsen - NADC
Dr. Joe Templeton - Texas A&M University
Dr. Fred Enright, Sue Hagius, Joel Walker and William Flahive - Louisiana State University

EVALUATION OF THE VACCINE EFFICACY OF RB51 ADMINISTERED ORALLY IN ELK

Philip H. Elzer[1], Gerhardt G. Schurig[2], Fred M. Enright[1], and Donald S. Davis[3].

[1]Department of Veterinary Science, Louisiana State University Agricultural Center, Baton Rouge, LA 70803; [2]College of Veterinary Medicine, Virginia Tech, Blacksburg, VA 24061; and [3]Department of Veterinary Pathobiology, Texas A&M University, College Station, TX 77843.

Wild ungulates are susceptible to the infection and disease known as brucellosis. *Brucella abortus* can infect elk (*Cervus elaphus canadensis*); and under experimental procedures, elk have transmitted the disease to cattle. There is circumstantial evidence that elk may have transmitted brucellosis to cattle under natural conditions. Large numbers of brucella-infected elk are found in the winter feedground areas of western Wyoming. The largest concentration of these elk frequent the winter feedgrounds of the National Elk Refuge, Jackson, Wyoming, which is administered by the U.S. Fish and Wildlife Service, Department of the Interior. To a lesser extent, brucella-infected elk also exist in Yellowstone National Park. Wild, free-ranging bison (*Bison bison*) are also known to harbor *B. abortus*. These animals continue to hamper the efforts of brucellosis eradication. Therefore the purpose of this study was to orally vaccinate elk with *B. abortus* strain RB51 to mimic oral vaccination of large numbers of animals on the winter feedgrounds.

Brucella abortus RB51 is a stable, rough variant of strain 2308 which is rifampin resistant and has been demonstrated to induce protection against virulent brucella challenge in swine, goats, mice, and cattle. RB51 does not produce any O-side chain antigens in its lipopolysaccharide; therefore animals vaccinated with RB51 do not illicit antibodies against the O-side chain. This is of great benefit in that all of the standard tests used to diagnosis brucellosis measure antibodies specific for the O-side chain. Animals vaccinated with RB51 will not make antibodies which react in the standard diagnostic tests thus "vaccine-induced titers" and "false positives" would not cloud eradication efforts.

A preliminary study using RB51 as an oral vaccine was first performed in cattle. Briefly, twenty brucella-naive crossbred heifers were divided into 2 groups. Group 1 (saline controls) received their normal ration of hay/pellets with karo syrup and saline poured over it; and group 2 (vaccinates) received their normal ration laced with $>10^{10}$ colony forming units (cfu) of RB51 resuspended in karo syrup and saline. All of the animals were individually housed during the vaccination and monitored until they ingested all of their rations. The animals were pasture bred; and at approximately 180 days gestation, they were challenged with 1×10^7 cfu of virulent B. abortus strain 2308. Strain 2308 has an established virulence in cattle, bison, and elk and typically induces abortions in infected animals. The delivery status of the animals was noted; and all calves or fetuses were necropsied immediately after birth or abortion with selected tissues bacteriologically examined. Uterine swabs and milk samples were taken from the cows and also cultured for brucella. Four weeks after parturition, the cattle were necropsied and tissues taken for culture. The results are as follows: the saline controls had 7/10 abortions whereas the oral vaccinates had 3/10 abortions; the challenge strain was recovered from 80% of the controls as compared to only 20% of the vaccinates. This study indicated that RB51 could be used as an oral vaccine and when administered by this route stimulated protective immunity against virulent challenge.

The oral exposure of elk with RB51 to protect against infection and abortion was investigated. Brucella-negative female elk were obtained from a site in South Dakota and transported to North Dakota for the oral vaccination study.

Female elk were pasture bred and orally exposed to RB51(or saline as a placebo) in December to mimic the feed ground situation as that would be the practical time and site of vaccination. The females were divided into two groups; group 1 received saline and group 2 received at least 10^{10} cfu of RB51 placed into their mouths following scarification with a float. At one month post-vaccination, 2 out of 26 vaccinates were blood culture positive for RB51. The pregnant elk were transported to Texas A&M University 8 weeks post-vaccination. At midgestation all of the elk were challenged conjunctivally with 1×10^7 cfu of virulent B. abortus strain 2308. The animals were monitored for abortions, weak, and live births. All of the calves were necropsied soon after birth, and tissues were collected for culture. The adult females were necropsied, and tissues were cultured for brucella.

Table 1. Vaccine efficacy of RB51 administered orally to female elk.

Delivery Status	
Nonvaccinated saline controls	RB51 vaccinates
9 abortions	5 abortions
2 stillborn	0 stillborn
4 weak (died)	0 weak (died)
0 live	4 live

Table 2. Culture status of elk females and calves following challenge with virulent Brucella abortus.

Culture Results	
Nonvaccinate saline controls	RB51 vaccinates
11/15 for strain 2308 (80%)	3/9 for strain 2308 (30%)

Conclusions

RB51 administered orally to female elk provided partial protection against virulent B. abortus challenge compared to non-vaccinated controls as demonstrated by abortion and colonization. Protection was measured by the number of aborted fetuses or calves which died after birth and the colonization of the fetuses and females with the challenge strain. In the non-vaccinated control group, 100% of the calves died and 80% of the animals were culture positive for strain 2308. However, in the RB51 vaccinated group, 55% of the calves died and 45% of them were born healthy; and only 30% of the animals were culture positive for strain 2308. Based on these findings, RB51 should be further investigated as a possible oral vaccine in elk which frequent feedgrounds.

Acknowledgments

We wish to acknowledge the following for assistance in this project:

Louisiana State University (Sue Hagius and Joel Walker), ND Governor's Office; Windcave National Park, Ross Rice; ND Elk Breeders Association; North American Elk Breeders Association; ND State Veterinarians, Robert Velure and William Rotenberger; Mitchell Charles; John Murphy; SD State Veterinarian, Daryl Thorpe; SD AVIC, Lynn Tesar; WY Game and Fish (Drs. Tom Thorne, Terry Kreeger, Beth Williams and Walt Cook); USDA/APHIS/VS (Drs. John Kopec and Mike Gilsdorf) and USDA/APHIS/VS Cooperative Agreements.

ISSUES IN VACCINATION FOR BRUCELLOSIS

Fred M. Enright, Department of Veterinary Science, Louisiana State
University Agricultural Center, Baton Rouge, LA 70803

WHAT MIGHT THE USE OF AN EFFICACIOUS VACCINE ACCOMPLISH?

WHAT WILL THE USE OF SUCH A VACCINE FAIL TO ACCOMPLISH?

Factors which modify the effectiveness of *any* brucellosis vaccine.

Host susceptibility
    –in herd differences based on genetics, sex, age
    –inverse relationship between degree of susceptibility and level of
      protection afforded by the vaccine

Duration of immunity
    –necessity for repeated vaccination

Level of exposure
    –most clearly demonstrated in the level of protection demonstrated
      by experimental challenge v. field challenge

What differences may be expected in field exposure under the following
    circumstances?
    –a dry lot packed with pregnant dairy cows
    –a 100-acre pasture containing 50 pregnant beef cows

A brief history of brucellosis vaccines and the development of Strain 19
    vaccine
    –initially sought a vaccine to prevent late-term abortions in cattle
    –use of virulent *B. abortus* isolates to infect (vaccinate) heifers prior to
      breeding
      a. partial protection against abortions
      b. cattle infected with these vaccines shed *B. abortus* in milk and
        could infect other cattle
    –1923-1924: Buck discovers Strain 19
    –by the early 1930s had demonstrated the effectiveness of S-19 in
      adults and calves

–1940s: S-19 became the official vaccine in the National Brucellosis
Program

TRADE-OFFS WITH S-19

–Initially used to vaccinate all females regardless of age
    Problems: vaccine infections
                    vaccine titers
–1950s and 1960s limited use to heifer calves
–by 1970s Dr. Nicoletti rediscovered adult vaccination
–1990s replacement of S-19 with RB51

The development of S-19 forced scientists to develop and/or modify how
S-19 was used.
–lower doses for adults
–lower vaccination age of calves
–different routes of vaccination

What did we learn?
–very young cattle not protected as well as adults
–still had vaccine titer problems
–oral vaccination/conjunctival vaccination yield better protection

Why live vaccines?

Vaccination of elk and bison with S19
–Elk reproduction
    –females become sexually mature by 2 years of age
    –males not actively involved in breeding until 3 years of age
    –breeding until 3 years of age
    –breeding season mid-September to mid-October
    –gestation 8.5 to 9 months
    –calving mid-May to mid-June

Brucellosis in elk
–brucellosis in elk causes abortions/premature deliveries
–50-70% of female elk that become infected with B. abortus lose their
    first calf following infection
–like cattle, most abortions occur during the last 1/3 of gestation

–retention of placentas and other forms of infertility associated with brucellosis in cattle do not occur in infected elk
–transmission of brucellosis from infected elk to susceptible cattle occurs
–experimental animals were closely confined and transmission was associated with delivery of infected elk calves
–transmission occurs during late winter and early spring

"Brucellosis transmission from elk to cattle is extremely unlikely to occur at any other time or circumstance, including normal calving in traditional elk calving ranges (Thorne et al., 1991)."

–elk normally seek remote secluded areas to calve
–elk have been incriminated epidemiologically in spreading brucellosis to at least 4 cattle herds adjacent to the GYA (circumstantial)

Response of elk to S-19
–27% of S-19 vaccinated elk aborted S-19 infected fetuses
–38% of 66 vaccinated female elk vs. 69% of 35 nonvaccinated lost their calves following challenge with virulent *B. abortus*

S-19 Vaccination of Elk

Recovery of 2308 at necropsy* of elk challenged 6 to 1 weeks post hand vaccination

	Number of animals	Number from which 2308 recovered
Vaccinates	11	6 (55%)
Controls	8	7 (88%)
Vaccinates	6	0 (0%)
Controls	2	2 (100%)
Vaccinates	12	7 (58%)
Controls	6	4 (66%)

* Necropsies performed 24 to 55 weeks post challenge.

Cumulative results of hand vaccinated elk challenged 6 to 10 weeks post vaccination

	Number of animals	Number from which 2308 recovered
Vaccinates	29	13 (45%)
Controls	16	13 (81%)

Recovery of 2308 at necropsy from hand vaccinated elk challenged 1 to 2 years post vaccination

	Number of animals	Number from which 2308 recovered
Vaccinates	16	7 (44%)
Controls	14	5 (36%)
Vaccinates	8	3 (37%)
Controls	4	2 (50%)

*Necropsies performed 27 to 37 weeks post challenge.

Cumulative results of hand vaccinated elk challenged 1 to 2 years post vaccination

	Number of animals	Number from which 2308 recovered
Vaccinates	24	10 (42%)
Controls	18	7 (38%)

These studies suggest that elk are more susceptible to B. abortus than cattle.

S-19 vaccination of adult elk is more effective than 2-19 vaccination of elk calves.

Experimental Infection of Bison in Their Second Trimester of Pregnancy

–Infected bison (10/12) aborted 47.5 days post challenge
–Infected cattle (11/12) aborted 69.2 days post challenge

This difference is most likely due to the levels of bacteremia. More *B. abortus* in the bloodstream at an earlier interval following challenge in bison than in cattle.

S-19 vaccination of adult bison (90-120 days of gestation)
  –hand injection with 5.3 x 10^8 cfu S-19 aborted 29 /48 (60%)
  –ballistic injection with 1.7 x 10^9 cfu S-19 aborted 34/44 (77%); nonvaccinated bison aborted 6/46 (15%)
  –63% (30-48) of the hand injected bison demonstrated seroreactivity 12 months post vaccination
  –80% (36/45) of the ballistically injected bison demonstrated seroreactivity 12 months post vaccination
  –2% (14/837) nonvaccinated contact bison sharing winter pastures with the vaccinates seroconverted
  –1 vaccinated cow was chronically infected with S-19 and aborted a second fetus 13 months post vaccination

These vaccinated adult bison were challenged with 1.0 x 10^7 cfu of *B. abortus* strain 2308 approximately 13 months post vaccination.

Protection against abortion	
Hand vaccinated	57% (16/28)
Ballistically vaccinated	79% (19/24)
Controls	4% (1/27)

Protection against infection	
Hand vaccinated	30% (9/30)
Ballistically vaccinated	44% (12/27)
Controls	0% (0/30)

Vaccination of Bison Calves with S-19

    –8-10 months old bison calves were vaccinated
    –these calves vaccinated by hand or ballistically or given saline only;
       were challenged with 1.0 x 10^7 cfu of B. *abortus* strain 2308 as
       pregnant adults about 2 years post vaccination

    Results
       –5% (5/96) S-19 vaccinated calves remained seropositive for 24
         months
       –S-19 vaccinated bison calves were *not* protected against either
         abortion or infection

    Conclusions
      1. Bison are very susceptible to B. *abortus* infection
        –an attenuated vaccine strain resulted in prolonged infection,
          persistent seroreactivity, abortions, and exposure of contact
          controls
      2. S-19 vaccinated adult bison demonstrated significant resistance to
        infection and protection from abortions
      3. S-19 vaccination of 8-10 month old bison calves failed to increase
        their resistance to infection or protect against abortion

While not satisfactory, the elk and bison S-19 studies suggest that:

    1. Adults are most effectively protected against infection.
    2. This partial protection against infection translates into increased
      protection from abortions–and may result in decreased exposure to
      other susceptible animals.
    3. Calfhood vaccination of bison is not effective.
    4. Calfhood vaccination of elk may not be effective–was the diminished
      protection due to the extended intervals (1-2 years) between
      vaccination and challenge or due to the inability to immunize elk
      calves?

With reference to *any* vaccine for brucellosis in wildlife–S-19, RB41, and
those not discovered:

Do the constraints related to vaccine safety imposed by its use in
    commercial livestock also apply to its use in wildlife?
Specifically, are vaccine-induced abortions ever acceptable in wildlife
    populations?
Perhaps the number of abortions induced in pregnant elk and bison by
    S-19 are excessive.
We have, however, already been shown that bison abortions due to S-19
    were adequate to expose and perhaps even immunize 2% of a
    nonvaccinated population of contact bison.
I now feel that an acceptable vaccine which *occasionally* causes abortions
    in vaccinated animals can serve to enhance immunity in
    nonvaccinated members of the herd.

Finally, I would like to answer the two questions stated in the beginning of
my presentation:

1. What will an effective vaccine accomplish?
   –It will limit the spread of virulent *B. abortus* in the population (herd)
   –It will enhance resistance to infection within the population (herd)
   –Thus, with time, numbers of new infections are reduced
2. What will such a vaccine fail to accomplish?
   –It will not eliminate the disease within the herd

I will end by quoting Dr. Paul Nicoletti:

"The control of brucellosis depends largely on two of the main
principles of disease management: prevention of exposure to susceptible
animals, and increasing resistance of the population through vaccination.
The best results are achieved through a combination of these, but
vaccination, especially in large cattle populations, is far more effective."

Oral RB51 Vaccination of Elk:
Tissue Colonization and Immune Response

A Study Conducted by the Wyoming Game and Fish Department
in Collaboration with
Louisiana State University and Virginia Polytechnic Institute

Methods

- 40 elk captured from the National Elk Refuge; card tested negative
- Transported to Sybille Wildlife Research Unit; re-tested several times
- 34 elk vaccinated
  —15 adults (10.5)
  —15 yearlings (7.8)
  —4 calves (2.2)
- 6 controls (6.0)
- Elk vaccinated orally with 6 X 10^9cfu *Brucella abortus*, strain RB51 every other day for three vaccinations
- Mucosa excoriated prior to vaccination
- Two elk necropsied every other week post vaccination for blood and tissue culture and histopathology
- Remaining elk bled every other week for serology and hemoculture

Results (preliminary)

- All elk remained serologically negative for field strain *Brucella abortus* (i.e., no vaccine crossreaction with standard tests)
- Last positive hemoculture, 54 days post vaccination
- Last positive tissue culture, 68 days post vaccination (3/43 tissues positive)
- Last detectable RB51 titer (ELISA 1:50), 10 weeks post vaccination

Conclusions (preliminary)

- Oral vaccination of elk resulted in tissue colonization
- Oral vaccination resulted in an immune response

RB51 Vaccination of Elk: Safety and Efficacy

A Study Conducted by the Wyoming Game and Fish Department
in Collaboration with University of Wyoming,
Louisiana State University and Virginia Polytechnic Institute

Methods

- 45 female elk calves captured from the National Elk Refuge in 1995; card tested negative
- Transported to Sybille Wildlife Research Unit; re-tested several times
- Elk vaccinated in May 1995
  - hand: 1×10^9 cfu (n=16)
  - biobullet: 1×10^8 (n-16)
- 13 controls
- Elk were bred in fall 1996
- All elk challenged in March 1997 with 1×10^7 cfu *Brucella abortus* strain 2308 intraconjuctivally
- Elk observed daily thereafter for abortion
- Elk necropsied after delivery/abortion

Results (preliminary)

- Hand: 14/16 aborted (88%)
- Biobullet: 12/16 aborted (75%)
- Controls 13/13 aborted (100%)

Conclusions (preliminary)

- There was no difference in abortion rate between vaccinates and controls ($P \geq 0.10$)

BRUCELLOSIS IN THE GREATER YELLOWSTONE AREA: WHAT IS THE PROBLEM?

Paul Nicoletti, D.V.M., M.S., Professor, College of Veterinary Medicine, University of Florida, Gainesville, FL 32611-0880

One would have to be the equivalent of Rip Van Winkle to not know of some of the controversies surrounding this subject. Passions and opinions are many and intense. Numerous articles have been published and the broadcast media has been busy and seen by millions.

The conflict between the natural wildlife and those who wish to protect them and those of private interests, especially of ranchers, is a classic example of a problem in the United States. Whether it is overpopulation of raccoons in Pinellas County in Florida, or too many deer on Long Island, or brucellosis in Yellowstone Park bison, resolution of these conflicts is difficult. There is a direct confrontation between the concepts of doing whatever is necessary to eradicate a disease and to leave natural forces to function.

My credibility to address this group is based upon a near lifetime career of specialization in brucellosis, an episode near Gainesville which also involved bison brucellosis, serving as an expert witness in trials and hearings and distance. Everyone knows that the further one is form the problem, the more expertise can be claimed.

Brucellosis is characterized in natural animal hosts by abortion, retained placenta, and pathologic lesions in males. The susceptibility to infection and severity among wildlife hosts have been studied and some results are conflicting. It is quite clear that under natural conditions, brucellosis in bison is of little consequence in fecundity. Bison are not shaggy cows and their behavior, physiology, and responses to infectious agents may be unique.

Control and hopeful eradication of brucellosis in domesticated livestock are based upon quarantine, vaccination, and slaughter of seropositive animals. Clearly, these methods are far more difficult to apply in wildlife hosts.

The USDA and states depend upon selected surveillance systems to identify possible infected herds of cattle as part of the national brucellosis eradication program. In beef cattle, this surveillance largely relies upon blood samples which are collected at slaughter. This system provides data upon which classification of states depends. Classification of states as "free" of cattle brucellosis allows more freedom in cattle commerce. The most affected states by the wildlife brucellosis issue of Wyoming, Idaho, and Montana are classified free. The threat of reducing this state status if one or more cattle herds become infected causes terror among ranchers. Some states have threatened boycotts of cattle movements. These scenarios are harsh and without scientific merit. Surely, disease control officials can have more wisdom in handling disease. The surveillance system which is used to detect a problem should also be evidence of the lack of a problem and there is still no evidence that bison of the YNP have been responsible for any transmission of brucellosis to area cattle. There is anecdotal evidence of transmission from wildlife to cattle in the National Elk Refuge feeding grounds area to a few herds. The elk are known to have a high prevalence of clinical infection and seropositivity.

It is important to understand the differences between seropositivity, infection, and disease. Many surveys among free-living bison have found a rather high seroprevalence of brucellosis. When specimens are examined bateriologically, only about 20 percent of those with antibodies are culture positive. Further, it is rare to isolate the bacteria from female reproductive organs. It is interesting that the highest percentage of culture positive bison is among young animals and bulls.

The migration and subsequent slaughter of over 1000 bison during the severe winter of 1996-1997 caused enormous outcries among many persons and groups. A further 600 or so starved within the park boundaries. The migration is apparently assisted by snowmobile paths and it has often been suggested that snowmobiles be banned from the park or use be restricted.

It seems epidemiologically correct to suggest that the elk feeding grounds in Wyoming present a far greater risk in disease prevalence and management than the bison of YNP. A project to vaccinate some of the elk with biobullets of strain 19 has been in progress for several years with reduction in seroprevalence among the elk.

Some Observations

1. It is hyperbole to suggest that if brucellosis cannot be eradicated from the GYA, that efforts to eradicate the disease from domestic animals have been wasted. Many believe that measures which would be necessary to eliminate brucellosis from the wild animals would eliminate the hosts.

2. The work eradication and the state classificaiton system must be modified to conform with reality.

3. The excessive attention to the bison and much less attention to the elk are driven by attitudes of ranchers towards the two species. Clearly, there must be some attitudinal changes or ranchers face possible eventual loss of privileges of using public lands for cattle grazing.

4. There is much agreement that the bison population within YNP needs more management. It remains very questionable if this should include possible vaccination to prevent brucellosis. There is no satisfactory vaccine, delivery system or evidence of a disease problem.

During testimony in the rather famous Parker lawsuit, Judge Bremmer asked "Dr. Nicoletti, what would you do with the problem of brucellosis in the Greater Yellowstone Area?" My reply was "Your honor, I don't know." While I have several observations and opinions, I feel that my answer puts me among a rather large company of others.

I appreciate the invitation to attend this conference and to present this paper. I trust that meetings such as this will educate and perhaps eventually, lead to some compromises and solutions to some very complex issues.

Safety and Efficacy of Existing Vaccines to Prevent Brucellosis in Bison

Steven Olsen, DVM, Ph.D.,United States Department of Agriculture, Agricultural Research Service, National Animal Disease Center, Ames, IA 50010

Protection and lasting immunity against brucellosis is achieved with vaccines containing live bacteria which stimulate a strong cell-mediated immune response. Factors enhancing cell-mediated immunity following administration of live vaccines may include prolonged antigenic stimulation due to proliferation of the vaccine strain within the host and internal antigen processing with more efficient presentation with major histocompatibility antigens (Class I) associated with cellular immune responses. Mouse models of brucellosis indicate that antibodies may have a minor role in short-term protection.[1,2] However, studies in cattle have demonstrated a poor correlation between the vigor of the humoral response and protection.[3] This is supported by data from cattle experiments in which vaccinated animals which were seronegative prior to midgestational challenge with a virulent *Brucella abortus* strain were protected against infection and abortion at a time of maximum susceptibility. Additionally, it is customary for animals which abort to have very high titers against brucellosis despite having failed to mount an effective immune response which prevented localization in placental and fetal tissues.

An ideal vaccine against brucellosis would persist long enough to induce good immunity without persisting into adulthood, would not cause clinical illness, and would not induce serologic responses which interfere with detection of animals infected with virulent field strains of *B. abortus*. Typically, vaccines against brucellosis are more efficacious in preventing abortions than preventing infection. Vaccination of wildlife with live vaccines would also have to consider potential detrimental effects on nontarget species, such as predators, which may inadvertently be infected with the vaccine strain.

Studies evaluating the safety and efficacy of brucellosis vaccines in bison are limited. When *B. abortus* strain 19 was administered by hand (1.7×10^9 colony-forming units (CFU)) or ballistic methods (7.7×10^9 CFU) to bison heifer calves, 5% of vaccinated calves had titers on brucellosis serologic tests at 2 years of age.[4] Following challenge with 1 x 10[7] CFU of virulent *B. abortus* strain 2308 during pregnancy, calves vaccinated with strain 19 averaged 25% abortions as compared to 30% abortions in heifers vaccinated with saline (controls). Only 9% of heifers calfhood vaccinated with strain 19 were protected against infection as compared to 17% of nonvaccinated controls. No statistical differences in abortion or infection rates were detected between bison calfhood vaccinated with strain 19 and nonvaccinated controls.

A second study evaluating strain 19 as a vaccine for adult bison indicated that a high percentage (58%) of pregnant animals aborted following vaccination.[5] When challenged with *B. abortus* strain 2308 during following pregnancy (13 months after vaccination), the percentage of abortions was less in strain 19-vaccinated bison as compared to nonvaccinated bison (33% versus 96%, respectively). In a similar manner, protection against infection was greater in strain 19 vaccinated bison as compared to nonvaccinates (39% versus 0%, respectively). In addition to its abortogenic effects, the strain 19 vaccine also induced persistent serologic titers on brucellosis surveillance tests and chronic infections in bison vaccinated as adults.

Research at the National Animal Disease Center has identified a new vaccine for cattle, *B. abortus* strain RB51, that is efficacious in preventing abortion and infection.[6,7] This vaccine does not induce antibody responses which cause positive responses on brucellosis surveillance tests[8,9] and therefore does not impair the identification of *Brucella*-infected cattle under field conditions. Research projects to evaluate strain RB51 as a vaccine for bison have been initiated at our facility.

A preliminary study to evaluate strain RB51 vaccination (10[10] CFU) of bison indicated that the vaccine is clinically safe in bison calves and does not induce positive responses on brucellosis surveillance tests.[10] Antibody responses against the vaccine strain were detected using a dot-blot test which has been demonstrated to have a high sensitivity and specificity in cattle.[11] Adverse clinical signs were not detected following vaccination of bison with strain RB51. The vaccine strain was still present at 16 weeks after vaccination in bison whereas cattle typically clear strain RB51 from the draining lymph by 12 to 14 weeks. These bison were raised to maturity and pasture bred. Data obtained following challenge at midgestation with 1 x 10[7] CFU of *B. abortus* strain 2308 suggested that strain RB51 induces some protection in bison. However, as nonvaccinated bison were not included in the challenge portion of the study, conclusions cannot be made on the efficacy of strain RB51 in bison without additional studies.

Additional studies have been completed evaluating calfhood vaccination of bison with 10^{10} CFU of strain RB51. These studies have provided further evidence that strain RB51 persists longer in bison when compared to cattle but does not appear to cause adverse clinical signs. Data from these studies suggests that strain RB51 localizes in lymphatic tissues and induces cell-mediated immune responses. Data from biosafety experiments have indicated that the strain RB51 vaccine is not shed from bison following vaccination.

The strain RB51 vaccine may have similar problems in adult bison as the strain 19 vaccine. When administered to pregnant bison at a 10^9 CFU dosage, strain RB51 appears to induced abortion in some animals.[12] This dosage is safe in pregnant cattle.[13] Ongoing studies will determine if adverse clinical or biosafety effects may limit the use of strain RB51 in adult bison bulls.

At the present time, the strain RB51 vaccine is the most likely candidate for use to prevent brucellosis in bison. Continued research efforts will be required to verify the efficacy of strain RB51 to prevent brucellosis in bison. Addition research will also be required to develop delivery methods and guidelines for the use of strain RB51 in management programs to reduce or eliminate *Brucella* infections in bison.

References

1. Winter, A.J., Duncan, J.R., Santisteban, C.G., Douglas, J.T., Adams, L.G. Capacity of passively administered antibody to prevent establishment of Brucella abortus infection in mice. Infect. Immun. 57: 3438-3444, 1989.

2. Cloeckaert, A., Jacques, I., Bosseray, N., Limet, J.N., Bowden, R., Dubray, G., Plommet, M. Protection conferred on mice by monoclonal antibodies directed against outer-membrane-protein antigens of Brucella. J. Med. Microbiol. 34: 175-180, 1991.

3. Nicoletti, P. Vaccination. In: Animal Brucellosis. K. Nielsen and J.R. Duncan eds. 1990. pp 283-300.

4. Davis, D.S., Templeton, J.W., Ficht, T.A. , Huber, J.D., Angus, R.D., Adams, L.G. *Brucella abortus* in bison. II. Evaluation of strain 19 vaccination of pregnant cows. J Wildlife Dis 1991; 27:258-264.

5. Davis, D.S. Summary of Bison/Brucellosis Research conducted at Texas A&M University 1985-1993. Proceedings of North American Public Bison Herds Symposium. July 27-29, 1993, Lacrosse, WI, pp 347-161.

6. Cheville, N.F., Stevens, M.G., Jensen, A.E., Tatum, F.M., Halling, S.M. Immune responses and protection against infection and abortion in cattle experimentally vaccinated with mutant strains of Brucella abortus.

Am. J. Vet. Res. 54: 1591-1597, 1993.

7. Cheville, N.F., Olsen, S.C., Jensen, A.E., Stevens, M.G., Palmer, M.V. Effects of age at vaccination on efficacy of *Brucella abortus* strain RB51 to protect cattle against brucellosis. Am. J. Vet Res 57: 1153-1156, 1996.

8. Stevens, M.G., Hennager, S.G., Olsen, S.C., Cheville, N.F.. Serologic responses in diagnostic tests for brucellosis in cattle vaccinated with *Brucella abortus* strain 19 or RB51. J. Clin. Microbiol. 32: 1065-1066, 1994.

9. Olsen, S.C., Evans, D., Hennager, S.G., Cheville, N.F., Stevens, M.G. Serologic Responses of Calfhood-Vaccinated Cattle to *Brucella abortus* strain RB51. J. Vet. Diagn. Invest. 8: 451-454, 1996.

10. Olsen, S.C., Cheville, N.F., Kunkle, R.A., Palmer, M.V., Jensen, A.E. Bacterial survival, lymph node changes, and immunologic responses of bison (*Bison bison*) vaccinated with *Brucella abortus* strain RB51. J. Wildlife Dis. 33: 146-151, 1997.

11. Olsen, S. C., Stevens, M.G., Cheville, N.F., Schurig, G. Experimental use of a dot-blot assay to measure serologic responses of cattle vaccinated with *Brucella abortus* strain RB51. J. Vet. Diagn. Invest. (In Press)

12. Palmer, M.V., Olsen, S.C., Jensen, A.E., Gilsdorf, M.J., Philo, L.M., Clarke, P.R., Cheville, N.F. Abortion and placentitis in pregnant bison (*Bison bison*) induced by the vaccine candidate *Brucella abortus* strain RB51. Am. J. Vet. Res. 57: 1604-1607, 1996.

13. Palmer, M.V., Olsen, S.C., Cheville, N.F. Safety and immunogenicity of Brucella abortus strain RB51 vaccine in pregnant cattle. Am. J. Vet. Res. 58: 472-477, 1997.

Lesions and Sites of Tissue Localization of
Brucella abortus in Female Bison from
Yellowstone National Park: Preliminary Results

Jack C. Rhyan,[1] Keith Aune,[2] Thomas J. Roffe,[3] Thomas Gidlewski,[1] Darla R. Ewalt,[1] and Michael Philo[4]

[1]U.S. Department of Agriculture, Animal and Plant Health Inspection Service, Veterinary Services, National Veterinary Services Laboratories, P.O. Box 844, Ames, IA 50010; [2]Montana Department of Fish Wildlife and Parks, Research and Technical Services Bureau, Montana State University Campus, Bozeman, MT 59717; [3]U.S. Department of the Interior, U.S. Geological Survey, Biological Resources Division, National Wildlife Health Center, Bozeman Station, Montana State University Campus, Bozeman, MT 59717; [4]U.S. Department of Agriculture, Animal and Plant Health Inspection Service, Veterinary Services, Western Region, 9439 Owl Way, Bozeman, MT 59715

Introduction

Brucella abortus produces abortions in cattle, bison (Davis et al., 1990; Rhyan et al., 1994; Williams et al., 1993) and elk (Thorne et al., 1978). Metritis and retained placentas have also been associated with the infection in cattle and bison (Corner and Connell, 1958; Williams et al., 1993). Seminal vesiculitis, orchitis, and epididymitis have been observed with *B. abortus* infection in male cattle and bison (Corner and Connell, 1958; Creech, 1930; Tunnicliff and Marsh, 1935; Williams et al., 1993; Rhyan et al., 1997). In a recent study, *B. abortus* was isolated from two or more tissues from six of seven young bison bulls that had recently seroconverted (Rhyan et al., 1997). The purpose of this study was to determine the most frequent sites of tissue localization of *B. abortus* in female bison from Yellowstone National Park (YNP).

Materials and Methods:

Between February 1995 and January 1997, specimens were collected from 26 seropositive adult female bison. Twenty-five of the animals were killed after leaving YNP, and one animal was killed by YNP personnel because it had a retained placenta and was in close proximity to the northern border of YNP. The cow had recently aborted as evidenced by the early date (March of 1995) and the lack of mammary gland development. No fetus or calf was found. Additionally, specimens were collected from a term fetus and placenta that were found near Gardiner, Montana, in April of 1996. Tissue specimens were collected from all animals for culture in accordance with the recommendations published by the Greater Yellowstone Interagency Brucellosis Committee (GYIBC, 1996). Additionally, portions of the uterus and placenta from the cow killed in YNP and portions of lung and placenta from the fetus found near Gardiner, Montana, were fixed in 10 percent neutral buffered formalin and routinely processed for histopathologic examination. Selected tissues were also stained using a previously described immunohistochemical technique (Rhyan et al., 1997) that employs a polyclonal antibody developed against *B. abortus* (Palmer et al., 1996).

Tissues were cultured using a previously described technique (Rhyan et al., 1997) in which each piece of tissue was individually minced, macerated with an equal volume of PBS in a stomacher, and further processed in a glass tissue grinder. The resulting slurry was then poured in aliquots onto the following media: tryptose agar with five percent bovine serum and antibodies (TSA), TSA with ethyl violet, Ewalts medium, and Farrel's medium. Plates were incubated with added CO_2 at 37 C for 2 weeks. Cultures were identified and biotyped using the techniques of Alton et al. 1988).

Sero status of the animals was initially determined using the card test and was confirmed with the following tests: standard plate, standard tube, rivanol, complement fixation (CF), buffered acidified plate antigen (BAPA), and particle concentrate fluorescence immunoassay (PCFIA). All animals chosen for this study were positive on multiple serologic tests.

Results:

At present, cultures have been completed on 16 of the adult bison and on the fetus. *Brucella abortus* was isolated from tissues of 7 of the 16 animals. The most common culture positive tissues were the supramammary lymph nodes (7/7), retropharyngeal lymph nodes (5/7), and iliac lymph nodes (5/7). *Brucella abortus* was isolated from 15 specimens including the placenta and feces from the bison with the retained placenta. The organism was also isolated from 15 sites cultured from the term fetus and placenta found near Gardiner. Histologically, lesions from both placentas and the fetus consisted of necropurulent placentitis and mild pleocellular bronchointerstitial pneumonia. Immunohistochemical staining revealed large numbers of brucellae in placental trophoblasts and in phagocytes present in placental and uterine exudate. Fetal lung also contained brucellar antigen in exudate in airways.

Discussion:

The preliminary results of this study suggest that the supramammary, iliac, and retropharyngeal lymph nodes are the most frequent sites of tissue localization of *B. abortus* in female bison from YNP. Additionally, the results from the cow that had recently aborted suggest widespread infection in that animal at the time of abortion. The presence of *B. abortus* in the feces probably resulted from ingestion of portions of the infected placenta and/or licking off the infected fetus. Similar findings in cattle have been reported. The placentitis and fetal pneumonia with large numbers of organisms in placental trophoblasts are consistent with lesions produced by *B. abortus* in cattle (Payne, 1959), goats (Meador et al., 1986), and captive bison (Davis et al, 1990).

REFERENCES

Alton GG, Jones LM, Angus RD, Verger JM: 1988, Techniques for the brucellosis laboratory. Institut National de la Recherche Agronomique, Paris, France, 190 pp.

Corner AH, Connell R: 1958, Brucellosis in bison, elk, and moose in Elk Island National Park, Alberta, Canada. Can J Comp Med 22:9-20.

Creech GT: 1930, *Brucella abortus* infection in a male bison. North Am Vet 11:35-36.

Davis DS, Templeton JW, Ficht TA, et al.: 1990, *Brucella abortus* in captive bison I. Serology, bacteriology, pathogenesis, and transmission to cattle. J Wildlife Dis 26:360-371.

Meador VP, Tabatabai LB, Hagemoser WA, Deyoe BL: 1986, Identification of *Brucella abortus* in formalin-fixed, paraffin-embedded tissues of cows, goats, and mice with an avidin-biotin-peroxidase complex immunoenzymatic staining technique. Am J Vet Res 47:2147-2150.

Palmer MV, Cheville NF, Tatum FM: 1996, Morphometric and histopathologic analysis of lymphoid depletion in murine spleens following infection with *Brucella abortus* strains 2308, RB51, or an htrA deletion mutant. Vet Pathol 33:282-289.

Payne JM: 1959, Pathogenesis of experimental brucellosis in the pregnant cow. J. Pathol Bacteriol 78:447-459.

Rhyan JC, Quinn WJ, Stackhouse LL, et al.: 1994, Abortion caused by *Brucella abortus* biovar 1 in a free-ranging bison (*Bison bison*) from Yellowstone National Park. J Wildlife Dis 30:445-446.

Rhyan, JC, Holland SD, Gidlewski T, et al.: 1997, Seminal vesiculitis and orchitis caused by *Brucella abortus* biovar 1 in young bison bulls from South Dakota. J Vet Diagn Invest 9: IN PRESS.

Thorne ET, Morton JK, Blunt FM, Dawson HA: 1978, Brucellosis in elk. II. Clinical effects and means of transmission as determined through artificial infections. J Wildl Dis 14:280-291.

Tunnicliff EA, Marsh H: 1935, Bang's disease in bison and elk in Yellowstone National Park and on the National Bison Range. J Am Vet Med Assoc 86:745-752.

Williams ES, Thorne ET, Anderson SL, Herriges JD Jr: 1993, Brucellosis in free-ranging bison (*Bison bison*) from Teton County, Wyoming. J Wildlife Dis 29:118-122.

APPENDIX C

OTHER DISEASES IN GYA WILDLIFE

Bacterial diseases other than *B. abortus* are present in the GYA and do infect bison and elk. Those and other species also can be affected by parasitic and viral diseases. It is useful to examine some aspects of these diseases with what is known of infection and transmission of brucellosis. As is the case with brucellosis, research and data are lacking in wildlife for many of the diseases discussed below.

BACTERIAL DISEASES

Tuberculosis

Tuberculosis caused by *Mycobacterium bovis* is a chronic bacterial disease that has tissue changes similar to those in brucellosis. Tubercular lesions develop in lungs and intestine, and transmission appears to occur by inhalation or by ingestion of contaminated material. Unlike brucellosis, lesions have not been found in the reproductive tract of bison or elk with tuberculosis, and placentae have not been shown to be infected. Although tuberculosis rarely has been diagnosed in free-ranging bison or cervids in North America, it is common in bison in Wood Buffalo National Park in Canada (Tessaro 1987). Tuberculosis has recently been reported in elk in Manitoba and in mule deer in south-central Montana (Rhyan et al. 1992). Disease and tissue lesions in asymptomatic animals are uncommon, and the risk of transmission of tuberculosis in bison and elk appears to be considerably lower than that of brucellosis in these species.

Previous reports of tuberculosis in free-ranging animals have been in white-tailed deer (*Odocoileus virginianus*) in New York, Michigan, and Ontario. In the Canadian National Buffalo Park near Wainwright, Alberta, gross lesions consistent with tuberculosis have been found in elk (*Cervus elaphus nelsoni*), mule deer (*Odocoileus hemionus*), and moose (*Alces alces*) (Tessaro 1987).

The presence of tuberculosis in captive herds of deer and elk in several states and provinces in North America might constitute a source of *M. bovis* for wild species. In each of the above cases, *M. bovis*-infected cattle, captive elk, or bison herds were in the vicinity and were considered likely sources of sporadic tubercular infections in the wild ungulates.

Recently, tuberculosis caused by M. *bovis* was diagnosed in an infected captive herd of elk near the northern border of YNP; the disease occurred near free-ranging northern YNP elk (Thoen et al. 1992). On the basis a single tuberculin skin test, the herd had 28 positive reactors; at necropsy, one animal had tuberculous lung lesions from which M. *bovis* was isolated. A followup disease survey of free-ranging, hunter-killed elk from three areas of YNP revealed no tubercular lung lesions in 289 elk collected between December 1991 and January 1993. Neither M. *bovis* nor M. *paratuberculosis* was cultured from specimens. Antibodies to B. *abortus* were found in serum samples from 0%, 1%, and 1% of elk from the three areas sampled (Rhyan et al. 1997).

If tuberculosis is suspect in bison or elk, the medial and lateral retropharyngeal, mediastinal, and tracheobronchial lymph nodes should be collected and examined bacteriologically and histologically. M. *avium* can cause tuberculosis in deer but is most often isolated from deer that have no lesions of tuberculosis (Rhyan et al. 1997).

Paratuberculosis

Paratuberculosis (Johne's disease), a chronic intestinal infection of cattle and other ruminants, is a progressive granulomatous enteritis that is seen clinically as severe diarrhea and wasting. Paratuberculosis has been reported in free-ranging ungulates, including bighorn sheep (*Ovis canadensis*), mountain goat (*Oreamnos americanus*), tule elk (*Cervus elaphus nannodes*), axis deer (*Axis axis*), and fallow deer (*Dama dama*). Paratuberculosis has been reported in red deer and has been reproduced experimentally in elk, mule deer, and white-tailed deer (Williams et al. 1983). The absence of clinical paratuberculosis and the negative culture results for M. *paratuberculosis* are consistent with the lack of reports on paratuberculosis in elk in national parks other than YNP (Rhyan et al. 1997). The risk of transmission of paratuberculosis in bison and elk appears to be low, although it does occur.

Leptospirosis

Leptospirosis affects the liver and kidney. Bacteria replicate in the renal tubules and are released into urine, and new animals are infected when they drink contaminated water. Serologic evidence of leptospirosis has been found in elk and in bison of YNP (Taylor et al. 1997). The mechanisms and risk of transmission in elk is not known. Abortion is associated with leptospirosis in most mammals, but the incidence of leptospiral abortion in elk and a role in transmission through genital infection are not known. In the southwestern United States, serologic evidence of leptospirosis suggests that deer are a natural host for leptospires.

Anthrax

Anthrax is acquired from ingestion or inhalation of bacterial spores in soil or on contaminated vegetation and debris; it is not transmitted directly from animal to animal and does not specifically involve the reproductive tract. Anthrax appears clinically as peracute septicemia in bison. Free-ranging bison with anthrax have been reported, and sporadic epizootics have occurred at various North American sites, including one outbreak in which 1,110 bison died. The causal organism, *Bacillus anthracis*, appears to be moved from endemic areas in Louisiana and Texas by waterfowl. In an outbreak in the Slave River lowlands and Wood Buffalo National Park, control was attempted with depopulation; 1,600 bison were killed (Broughton 1987). Transmission from bison to cattle has not been reported, but human infections from bison anthrax have been reported in several areas (Tessaro 1989). A bison-vaccination program was initiated in 1965 in Canada, but it was discontinued in 1978.

Other Granulomatous Bacterial Diseases

Yersiniosis

Disseminated microabscesses surrounding colonies of *Yersinia pseudotuberculosis* occur sporadically in many species of wild mammals and birds. Wild rodents are reservoirs for this bacterium, and ingestion of grass contaminated with feces and predation by carnivores are sources of infection. Epizootics of yersiniosis have been reported in farmed cervids, including elk, fallow deer, red deer, and red-deer/elk hybrids (Sandford 1995). Yersiniosis is not an important disease in the GYA, and, although antigens of *Yersinia* spp. are known to cross-react with those of *Brucella* spp., there is no evidence that this is important in serology of bison and elk. Placentitis and abortion caused by *Y. pseudotuberculosis* occur in domestic sheep and goats, but transmission has not been associated with the reproductive tract. Serologic studies have shown that 86% of adult free-ranging YNP coyotes are seropositive for *Y. pestis* (Gese et al. 1997).

Pasteurellosis

Pasteurella multocida causes respiratory disease and septicemia in elk, and those diseases have been documented in YNP and in the NER (Franson and Smith 1988). The risk of transmission and mechanism of infection of *Pasteurella* spp. infections in bison and elk are not known. Granulomatous lesions resembling actinobacillosis lesions have been reported in lymph nodes of elk in the northern YNP region. Consisting of aggregates of macrophages with dense "sulfur granules" composed of debris and bacteria, they can be confused with tuberculosis. A recent survey in YNP found an incidence of 15%. In some cases, *Pasteurella hemolytica* has been isolated from affected tissues. These lesions appear to be transmitted by contamination of wounds with the bacterium and might also be caused by different species of bacteria (J. Rhyan, APHIS, pers. commun., 1997).

Vulgovaginitis

Chronic inflammatory pyogranulomatous mucocutaneous lesions of the vulva are common in elk. The lesions can be large, ulcerating, and persistent. It is thought to be caused by an organism that resembles *Corynebacterium renale*. The pathogenesis has not been established. Whether vulgovaginitis interferes with reproduction has not been reported.

Parasitic Diseases

Lungworms

Lungworms, *Dictyocaulus* spp., are common in elk in YNP; their incidence increases in the spring. The parasites are identified as *D. viviparus* elk strain or *D. hadweni.* In the 1960s, dissections of lungs of 59 YNP elk revealed lungworms in five animals. A study in Teton County, Wyoming, found incidences of 8%, 19%, and 15% in elk. Lungworms in land mammals are not associated with brucellosis, but in the new emerging forms of brucellosis in marine mammals, lungworms have been shown to carry *Brucella* spp. (Garner et al. 1997).

Ostertagiasis

Ostertagia ostertagi, a parasite of cattle, also infects bison. Bison-to-cattle spread has not been studied, although it has been stated that the capacity of bison parasites to infect cattle is of concern (Marley et al. 1995).

Scabies

Scabies is a highly contagious, enzootic infestation of wild ruminants and has been a problem for GYA elk. Rates of transmission are probably high, but scabies is a self-limited disease and does not typically cause debility or death. *Psoroptes* spp. burrows into the superficial layers of the skin to cause extensive chronic inflammation.

Viral Diseases

Few viral diseases are viewed as major causes of morbidity and mortality in bison and elk of the GYA, and data on risk of transmission are inadequate. Systemic viral infections similar to infectious bovine rhinotracheitis and bovine viral diarrhea exist, or could exist, in bison or elk. Serologic evidence of infection with bluetongue, epidemic hemorrhagic disease, infectious bovine rhinotracheitis, and bovine viral diarrhea can be found in elk and are most likely a reflection of their contact with cattle. Bluetongue and epidemic hemorrhagic disease can be lethal in deer species.

Chronic wasting disease (CWD) of elk and deer is an infectious nervous system affliction that resembles scrapie in sheep in its clinical signs, distribution of lesions in brain, and presence of scrapie-associated prion protein in affected tissue (Spraker et al. 1997). Surveys of hunter-killed animals within a 100-mile radius of Fort Collins, Colorado, and Laramie, Wyoming, have shown a 6% incidence in mule deer and 1% incidence in elk (T. Spraker, pers. commun., 1997). Evidence of the etiologic agents is found in brain, spinal cord, and lymph nodes (by using a monoclonal antibody derived from antigens of ovine scrapie). CWD is considered one of the transmissible spongiform encephalopathies that are potentially transmissible to other species, including humans. The disease has not been reported in elk or deer in the GYA, and no studies have been done on the danger that this disease has for humans.